ENGAGED
DESIGNING FOR BEHAVIOR (

Amy Bucher

 Rosenfeld®

NEW YORK   2020

"*Engaged* draws from the latest science on behavior change and compelling examples to teach design professionals to create world-class products that make big impacts on users' lives. Clearly written and well-researched, *Engaged* is a great addition to any designer's toolkit."

—Scott Sonenshein,
Rice University professor and bestselling author of
Stretch and *Joy at Work*

"An incredibly comprehensive and intelligent guidebook for designing products and services that change people's lives."

—Nir Eyal,
Bestselling author of *Hooked* and *Indistractable*

"A book is only as good as the behavior it changes, and in this practical guide, Bucher focuses on combining rigorous application with an approachable voice. The product of over a decade of experience across a variety of domains, *Engaged* is a unique contribution that is well worth reading for anyone who wants to create change in the world, particularly in a digital medium."

—Matt Wallaert,
Author of *Start at the End:
How to Build Products That Create Change*

"Amy's book is shockingly practical, showcases the impact of behaviors on design in practice, and provides clear tips and approaches you can immediately employ to benefit your work."

—Alyssa Boehm,
User experience executive

"Bucher proves in this step-by-step guide that behavior change design is valuable to all areas of design."

—Lis Pardi,
Experience Design Manager, Toast

"Interest in behavioral economics has exploded in recent years as product teams strive to empower people to achieve meaningful goals. For all on that journey—entrepreneurs, intrapreneurs, product managers, designers, researchers, engineers—this is a comprehensive playbook that translates behavioral economic theory into practical, actionable, and ethical design techniques."

—Jen Cardello,
VP, Head of UX Research & Insights, Fidelity Investments

"I have witnessed large audiences of people be mesmerized by Amy Bucher's presentations on digital health and behavior change design. She makes the topic approachable bringing academic theory to life through relevant examples and actionable insights. This book proves to be an essential guide for digital health innovation, moving beyond shiny objects to define what will truly deliver engagement and improved outcomes."

—Amy Heymans,
Co-founder and chief experience officer of Mad*Pow

"This book demystifies the psychology behind behavior change and offers practical methods and examples for applying it to product design. A worthwhile introduction for anyone trying to help users achieve health, financial, or other goals."

— Kim Goodwin,
Author of *Designing for the Digital Age*

Engaged
Designing for Behavior Change
By Amy Bucher

Rosenfeld Media LLC

125 Maiden Lane, Suite 209

New York, New York 10038

USA

On the Web: www.rosenfeldmedia.com

Please send errors to: errata@rosenfeldmedia.com

Publisher: Louis Rosenfeld

Managing Editor: Marta Justak

Interior Layout Tech: Danielle Foster

Cover Design: The Heads of State

Indexer: Marilyn Augst

Proofreader: Sue Boshers

For Bob and Melissa
My OGs

HOW TO USE THIS BOOK

Who Should Read This Book

This book is for anyone who wants to apply behavior change science to the design and development of digital products. Whether you're a social scientist working to change people's behaviors through apps, websites, and other digital tools, or a product manager, researcher, or designer who wants to infuse behavior science into your work, this book is for you.

What's in This Book

Chapter 1, "A Kind of Magic: Psychology and Design Belong Together," introduces the concept of behavior change design and identifies some of the domains in which it is used. You'll learn about the self-determination theory of motivation, which provides the underlying structure for the rest of the book. According to self-determination theory, the most long-lasting motivation comes when people's needs for autonomy, competence, and relatedness are supported.

Then, in Chapter 2, "Pictures of Success: Measurement and Monitoring," you'll learn about measuring and monitoring to ensure that your product is on track to achieve desired outcomes. Although most people think of outcomes assessment as something that happens late in the process, the seeds for a successful measurement strategy are planted at the very start of a behavior change design project.

The next chapters of the book talk about ways to support autonomy through digital design. Chapter 3, "It's My Life: Making Meaningful Choices," is about offering users meaningful choices as they approach behavior change, while Chapter 4, "Weapon of Choice: Make Decisions Easier," focuses on making choices easier for people so that they can end up on the right path more quickly.

The next set of chapters focus on supporting competence. In Chapter 5, "Something in the Way: Diagnosing Ability Blockers," you'll learn about using research to identify the things that block and boost users' ability to achieve their goals. Chapter 6, "Fix You: Solving

Ability Blockers," is about linking those findings to features that help users overcome obstacles. And in Chapter 7, "Harder, Better, Faster, Stronger: Designing for Growth," you'll learn about structuring goals, milestones, and feedback to keep users moving toward success.

Relatedness can be supported through connection. In Chapter 8, "Come Together: Design for Connection," I share how digital products can facilitate relationships between people. Then, in Chapter 9, "Mr. Roboto: Connecting with Technology," you'll see how technology can be used to help people feel a sense of connection. In Chapter 10, "A Matter of Trust: Design Users Can Believe In," I talk about trust, an essential ingredient in a healthy relationship between your user and your product.

Then, Chapter 11, "Someday Never Comes: Design for the Future Self," provides an overview of how the techniques from the previous chapters can be used to support a major behavior change challenge: getting people to do hard things today for the benefit of their future selves. Finally, in Chapter 12, "Nothing's Gonna Stop Us Now: Go Forth and Engage," I share how to carry behavior change design forward into your organization and your work.

Throughout the book you'll meet other experts working in behavior change design in sections I've called *Perspective*. These are people who have deep expertise in the topic of a chapter, and can either offer a deep dive or pro tips to help you as you learn. They are also all people whose work I admire, and many of them have influenced the way I practice my craft.

What Comes with This Book?

This book's companion website (rosenfeldmedia.com/books /engaged-designing-for-behavior-change/) contains a blog and additional content. The book's diagrams and other illustrations are available under a Creative Commons license (when possible) for you to download and include in your own presentations. You can find these on Flickr at www.flickr.com/photos/rosenfeldmedia/sets/.

FREQUENTLY ASKED QUESTIONS

I've got a background in behavior science, but no talent for visual design. Can I do behavior change design?

Absolutely. I was a total amateur at all of the things I thought were design before I started working in the field (and am still not very good at many of them). My strengths are research, strategy, and evaluation, so I partner with people who bring the visual and interaction design and application development chops. I have colleagues who have stronger design skills and less research experience, so they team up accordingly. It's all about building a team that can complement each other. Chapter 12 offers tips for bringing behavior change into your work, regardless of your background.

Is behavior change design actually necessary?

If I didn't think behavior change design was important, then there'd have to be something deeply wrong with me to have spent this much time and effort writing a book about it. Behavior change design helps make products more engaging, which means more people want to use them. That's good for business. And if your product is actually trying to change people's behavior, which is true of most products in industries like health, education, and sustainability, then behavior change design will hugely increase the odds it works. Learn more about how to measure the effects of behavior change design in Chapter 2.

What role does social media play in behavior change design?

Social support can play an important role in helping people change their behavior, and social media can deliver that support in a scalable way. But like any tool, social media must be used thoughtfully to produce the best results. Chapters 8 and 9 cover how to facilitate social support within behavior change design, both with and without connecting people directly to each other.

This book is mostly about motivation psychology. Are there other types of psychology that designers should learn?

Yes! Cognitive psychology is full of useful information for designers, especially visual and interaction designers and anyone creating content. This information includes how people perceive information and can guide decisions about how to present and format the flow of your product. Behavioral economics, which is psychology-adjacent, is what a lot of people think of when they think of behavior change. It's worth really understanding what behavioral economics is and is not.

Beyond that, read widely and often. Many of the topics that get covered in pop psychology don't fit neatly within a particular theory, but are helpful in thinking about designing for behavior change. See Chapter 12 for more suggestions on continuing your behavior change design education.

Can I use behavior change design for evil?

Sure, but I don't condone it, and it will probably come back to bite you when people realize what's happening and stop trusting you. Play the long game and use your behavior change design skills for good. Learn about how to build and maintain user trust in Chapter 10, and get tips from experts on ethical design practices in Chapters 6 and 10.

CONTENTS

FOREWORD

Behavioral science is having a moment. Nobel prizes, bestselling books, sought-after experts, speakers and consultants. People of every station peer over at the field, wondering if it might be a fix for their personal, professional, and community challenges. The field, which explains the hidden forces that drive human decision-making, is becoming a vital tool for building the future.

Amy Bucher jumps headfirst into this moment with *Engaged*, a practical, fun, and inspiring book about designing for change. Dr. Bucher knows that behavioral science doesn't provide a silver bullet to solve our problems in a single shot. So she outfits us with a bandolier full of bullets, explains what each does and why, shows how others have fired them, gives a little moral guidance, and sets us off, like Rambo, to save the day.

Okay, that dated reference was a little dramatic. The point is: I love this book and everything it offers us.

It's an accessible, relatable, and usable real-world how-to that should be on every designer's nightstand. Not only does Dr. Bucher explain the science for both newbies and seasoned pros—she makes big, scary words seem small and friendly—but she also uses hundreds of specific examples that pull the concepts right off the page. Reading about designers who have already executed these ideas helps us picture ourselves doing the same. She finishes each section with perspectives from leading practitioners, which not only brings our challenges and opportunities into greater focus, but also makes me wonder why I was relegated to the foreword. *What the heck, Amy?*

Despite that slight, what I really love about *Engaged* is that it's grounded in science. Dr. Bucher put the science first, even if that costs her some easy "solutions." She knows that behavioral science tempts some people either to sell easy fixes where none exist or to dismiss it all as some "cheesy Instagram motivational quote rabbit hole." In order to avoid either of those fates, Dr. Bucher calls on us to be guided both by the scientific method and by ethics. We can stick to clear ethical, moral, and trust-based principles and still make life-changing products.

Oh, and this dang thing is well-written, too! Dr. Bucher packs a lot of ideas in here, but does so with depth and appreciation for our cognitive skills and limits. She does it, I think, not so we just throw these ideas against the wall, but so that we appreciate the holistic, well-rounded, contextual approach, which will be key to the design of the future. Yes, our bandolier has a lot of bullets, but Amy loads each one with care, understanding, and purpose.

People will love this book because, like an 80s stand-up straightening his or her necktie, we're all asking, "What's the deal with behavioral science?" How can we balance optimism, progress, and excitement with pragmatism and a dash of caution? What is the guidance and knowledge we need to ensure that we're riding this moment in the right direction? Amy Bucher's *Engaged* answers those questions.

I wish I had written this book, and I can't wait to share it with the world.

—Jeff Kreisler
Editor-in-chief, PeopleScience.com
Coauthor, *Dollars and Sense:*
How We Misthink Money and How to Spend Smarter

INTRODUCTION

When I tell people I'm a psychologist, there are certain reactions I've come to expect. One is people asking if I can read their minds. (No. That's a psychic.) Another is people asking if I am diagnosing them based on our conversation. (No. I'm not a clinician, and even if I were, why would I work for free?) And very often, people ask if I'm able to force others into behaving a certain way using the tricks of psychology. (Another no.)

For better or for worse, psychology is not magic. And it's certainly not about forcing people to do something! There's no guarantee that using psychology will yield the desired outcome 100% of the time for 100% of the people. But what psychology *can* do is increase the odds that products built with its tools and insights will be effective at changing specific behaviors in the people who use them. Psychology may not be magic, but when I was a newly minted Ph.D. first applying its tools to design digital experiences that made people healthier, it sure felt like it.

Psychology offers both scientific tools and methods that can be used to understand what influences behaviors, as well as evidence-based techniques to change them. I'm excited to share how you can use psychology to make the digital programs you design and build more engaging for users. If your products are intended to change people's behavior, then psychology is essential for your design toolkit. Even if they're not, a strategic dash of science will still help you create better products.

Aside from genuinely geeking out about using psychology as a design tool, I also wanted to write this book to help people join me in this type of work. Over the years, I've met so many people who wanted to become behavior change designers but don't know how. There aren't many formal training programs (yet), and there is no curriculum of essential knowledge. And while there are a lot of great behavior change frameworks, tools, and books, there aren't very many that specifically apply the psychology of motivation to the design of digital experiences, the thing that's been the core of my job for more than ten years. I've done lots of coffee meetings and phone calls and blog posts, but I wanted a more enduring resource that people could really use. This book, I hope, will be it.

Relax, Take It Easy

While this book has mostly been a delight to write, it's also incited a few existential crises and more than one extended bout of procrastination. It was during one of those latter phases that I started combing Spotify for song titles related to the content of the book. Please enjoy the chapter titles and send me any suggestions I should add to the book's unofficial soundtrack.

With a Little Help from My Friends

Every chapter concludes with an interview with someone who is an expert in the topic of that chapter. I wanted to include other voices alongside mine to show the variety of ways that behavior change design could be done. In some cases, I knew that an expert interviewee could provide an in-depth example or specific pro tactics that would enrich the main content of the chapter. In others, the experts provided a wider lens that helped the reader view the chapter in a broader context.

These interviews were such a treat to do. They were an excuse for me to connect with old friends and make new ones, and to highlight people whose work has inspired me. I'm grateful to all of them. I hope you'll think they're as fantastic as I do.

Truth Hurts

Over the last decade or so, the field of psychology has entered a "replication crisis." Basically, when researchers repeated classic psychology studies, they got different results, putting the initial findings into question. In some unfortunate cases, it seems the original researchers fudged their results. Those studies are as good as dead.

But other cases are less clear-cut. Maybe the statistical analyses were not as stringent as they should have been, so the effect exists but is weaker than originally thought. In some older papers, the original methods weren't described well enough to follow exactly, or are no longer appropriate, given advancements in technology and knowledge. (Try doing a study on people who are new Internet users—it's near impossible now.) Or sometimes, some studies on a topic replicate and others don't. In these middle cases, more research is needed to figure out what's really going on.

I mention this because some of the topics in this book have been touched by the replication crisis, such as willpower and the growth mindset. I carefully reviewed the current state of the evidence for the studies I included and feel comfortable that they're valid. I also omitted some classic studies, with no small heartache, because the science is still in question.

The replication crisis means there's a little extra vigilance needed from anyone who uses psychological research in their work. Incorporating a literature review into your behavior change design process as an early step will help you avoid leaning on outdated studies.

Another Brick in the Wall, Pt. 2

One thing I get asked a lot, probably because I have one, is if you need a Ph.D. to do behavior change design. The answer is no. Some of the most talented behavior change designers I know do not have a Ph.D.; in fact, some of them didn't even study psychology in school. Work experience, reading, and on-the-job training can all whip someone into behavior change design shape over time. On the flip side, people like me with strong behavior science knowledge and research skills likely need the same kind of experience-based training to develop UX and design muscles. No one comes to behavior change design perfectly formed.

Whatever skills and background you bring to your work, I hope this book will add some new behavior change design tools to your repertoire. Take the things you learn, try them out, measure your results, and keep iterating until your designs do what you want them to do. Along the way, you'll infuse your own style and ultimately develop your own tried-and-true flavor of behavior change design. I can't wait to see it.

CHAPTER 1

A Kind of Magic

Psychology and Design Belong Together

In 2018, a team of researchers made headlines with the findings in their study of workplace wellness initiatives. The story that got me to click was titled "Study Finds Virtually Zero Benefit from Workplace Wellness Program in 1st Year."

This headline is alarming on its face. Workplace wellness is a multibillion dollar industry in the United States. If it doesn't work, that's a lot of money and time wasted. More alarming for me personally, it's where much of my professional work has focused in the past fifteen years. Was I tilting at windmills the whole time?

My nerd powers on alert, I downloaded the original research study that prompted the headlines. The Illinois Workplace Wellness Study is a multiyear examination of wellness programs for employees of the University of Illinois at Urbana-Champaign. At the end of the first two years of the study, the researchers found that although more people did health screenings once the program was put in place, it didn't seem to have any effect on medical spending, health behaviors, productivity at work, or health status. Unfortunately, those non-outcomes are exactly what most workplace wellness programs are supposed to improve, so these results do look pretty bad for workplace wellness.

But here's the thing. As I read the description of what the workplace wellness program at the University of Illinois actually consisted of, it became clear that *of course* it wasn't having its intended effects. The program, called *iThrive*, consisted of annual biometric screenings, an annual survey called a *Health Risk Assessment (HRA)*, and weekly wellness activities. Employees were given financial incentives for completing the screenings and HRA each year, and given paid time off to do the weekly activities, most of which were in-person classes. There was also an "online, self-paced wellness challenge"; however, no description was offered of how it was designed or what it included. Employees were encouraged to choose activities related to their HRA results. All of these features were pretty standard for workplace wellness programs, based on the information given, but none of them were designed for engagement.

If the program designers had adopted a behavior change design perspective, similar to the process this book lays out, iThrive probably would have looked a little different. For starters, most HRAs don't provide feedback that would help people choose the right behavior change programs; a program built on behavior change would provide more structured guidance to match people to goals.

Then there was a lack of variety in the programs offered, with an emphasis on group classes. People who hate group activities were highly unlikely to enroll in one, even for a reward. And about those rewards: research suggests that linking repeated behaviors, like new health habits, to financial rewards is a great way to make sure that people *don't* develop an intrinsic interest in doing them. Also, participation in the iThrive program was tethered to the workplace, which made it hard for people who valued keeping their health private from colleagues to participate without feeling uncomfortable. Finally, I don't even know what was in the digital component of the program, but chances are, it represented lots of missed opportunities to engage users in a wellness process.

It's not that workplace wellness programs can't change behavior. It's that workplace wellness programs are designed and implemented without a firm basis in psychology, so they don't work effectively for the way that human beings actually behave.

Behavior change design as a discipline can help prevent headlines like the ones about the Illinois Workplace Wellness Program by helping designers create more engaging, effective programs. Behavior change design offers a toolkit to build products that actually work, while also supplying the evidence to prove it. Specifically, behavior change design includes:

- A *process* for designing and building products that incorporate research and evidence
- Access to *frameworks and theories* to help leverage proven techniques within products
- *Tools* to define and track product success metrics

Working within a discipline like behavior change design helps ward off huge investments in programs that don't engage users or produce results for customers. The common language and process it offers sets expectations for potential investors and buyers that helps them assess whether a product is worth paying attention to. It also helps the product team do their work with rigor, detect and address potential problems early, and collect the evidence to either prove their worth or send them back to the drawing board. It's not a panacea, but having a well-defined method sure helps keep people honest.

So what is this method? Behavior change design is the application of psychological methods and research to the development of products, services, or experiences. Almost everything designers make has some behavior change built into it. Any time you expect a person to interact with your product, you're asking them to change their behavior from what it would be if the product didn't exist. The complexity, longevity, and significance of those behaviors can vary widely. As each of those dimensions increases, the need to include formal behavior change considerations in your design does too.

What Types of Products Benefit from Behavior Change?

Some products, services, or experiences are intended to change people's behaviors in the real world. Behavior change designers call those products *interventions*. That sounds very clinical, but an intervention doesn't have to be dry or complicated. Some of the behavior change interventions you've probably heard of include MyFitnessPal, Runkeeper, and Duolingo, none of which feels like a heavy experience. But all three of them get people to do something differently on purpose: MyFitnessPal encourages users to be mindful of their eating and movement; Runkeeper helps people train to run longer or faster; and Duolingo teaches new ways of communicating.

Behavior change interventions are more common in certain subject areas. In this book, you'll find a lot of health examples, because that's the domain I know best from my career and one where behavior change has been embraced. Health interventions may help people along a spectrum of functioning—from coping with acute illness or injury to managing chronic conditions, supporting wellness, or reaching sports and performance goals. And they can target a range of behaviors, including eating, exercise, taking medication, going to doctor's appointments, or deep breathing through stressful situations.

Behavior change interventions in financial services may center on major life goals like going to college (and paying off the associated loans), buying a home, or saving for retirement. Some successful financial behavior change interventions include changes to tax notices to prompt timely payment and changing 401(k) enrollment processes so that more people sign up.

Education is a natural outlet for behavior change; if people are building deep knowledge or new skills, they'll need to engage in practice behaviors. Some types of education manifest through behavior, like speaking a new language, writing code, or repairing an automobile.

The performance management tools that big companies use to review employees and manage bonuses are a type of behavior change intervention.

Environmental science organizations practice behavior change, too, whether it's getting people to consume fewer plastics or choose more sustainable fish to eat. As people realize the impact that their individual behaviors might have on the global climate, more digital interventions are being developed to support them in changing their efforts.

NOTE WHAT BEHAVIOR CHANGE MEANS TO PEOPLE

Early in the process of writing this book, I asked people on Twitter to recommend behavior change apps they had used. The overwhelming majority were workplace wellness apps, the type of health interventions you might get as part of your employer-offered health insurance. Even accounting for the fact that a lot of my Twitter followers are in the healthcare industry, it's striking that health is the first thing that comes to mind when people think behavior change.

Behavior change design can also be used to make consumer products and experiences more engaging—while sometimes having the positive side effect of helping users develop new habits or skills. Pokémon Go! is an example. It was designed as a game, but users report boosts in their daily step counts as a result of their quests to capture Pokémon. Even products without much potential for positively changing people's behavior, like shopping websites or music apps, could be made stickier using the strategies you'll learn about in this book. In fact, many of the most "addicting" digital experiences borrow heavily from psychology in their design.

But just because you can do something doesn't mean you should. It's easy to creep into dark patterns and manipulative design choices, if your goal in applying psychology is to keep someone within your product as long as possible without it being beneficial to them.[1] Behavior change design is about helping people achieve *their* goals, not yours.

---

1 Case in point, the gambling industry cleverly employs dark patterns to keep people glued to slot machines.

In this book, I focus on digital products—apps, websites, connected devices, and the ways in which they intersect. Many behavior change endeavors today include a digital component or are entirely digital; technology makes interventions scalable, so they can be delivered quickly and cheaply to large groups of people no matter where they live. And digital offers opportunities to reach people in or near moments where they're taking actions that matter. It's a channel with enormous promise for affecting outcomes.

Like it or not, people are going to use tactics from psychology to make their digital products more engaging. They might as well learn to do it right.

Where Does Behavior Change Happen?

Behavior change design works at two levels. For products that are intended to change people's behaviors, there is often a protocol built into the product itself. These protocols are step-by-step processes that outline the correct way to change a behavior based on previous research. For example, research on smoking cessation clearly indicates that setting a quit date in advance makes people much more successful at quitting, so most smoking cessation programs include steps around setting a quit date. Behavior change designers may be responsible for developing the protocol within a product, often in partnership with subject matter experts like physicians or researchers. Or, they may need to translate a protocol that exists in a nondigital format to a digital one; you'll see examples in this book where techniques like cognitive behavioral therapy (CBT), typically used in counseling, are brought into a digital experience. Creating or translating these sorts of protocols requires understanding their active ingredients and being able to make sound judgments about how to represent them accurately through digital experiences.

The second level at which behavior change design works is making the digital product itself engaging by aligning it with people's motivational needs. It is this second level of behavior change design that can be applied to nonbehavior change products, and where most of the material in this book is focused.

Although I'll primarily talk about using behavior change within the guts of a digital product to make it engaging, effective engagement also requires you to pay attention to the context in which the product is being used. That includes how your product is marketed and distributed, any reminders or messages users might receive from the product, and how data is collected about users' experiences. Some digital products include an onboarding experience with physical world components, for example, if there are connected devices that need to be set up. Others are designed to facilitate real-world conversations; consider someone with a health condition sharing their medication data from the app with a doctor during an appointment. Designing with an eye to how those experiences unfold will support engagement within the digital product itself.

Of course, because most behavior change takes place off the screen, behavior change designers must understand users in their real-life contexts, beyond their use of the product itself. The research that goes into understanding users and their needs almost always extends into the analog world. When the goal is to change something offline, the digital product becomes a tool rather than an end in itself.

> **NOTE** DON'T MAKE ASSUMPTIONS
>
> Understanding how behavior unfolds in the real world is crucial to design to support or change it digitally. In reviewing apps that include behavior change elements for this book, I've been (perhaps naively) surprised by how often stereotypes about user groups get baked in. Two things I noticed over and over: weight loss is congratulated, even when the product can't possibly know if it was intended or wanted. And products that asked about sexual activity presumed that the partners were male and female. Both of these assumptions could be really off-putting to a potential user who doesn't fit them. There are many ways to build flexibility into your product to avoid these embarrassing gaffes. Doing good research up front will help you recognize where you need them.

That said, behavior change design is a business. Most products include business goals that live alongside the behavior change goals. Products may carry subscription fees, require users to pay for access to premium features, or urge them to purchase expensive connected devices for enhanced functionality. Behavior change design can be an excellent tool to keep people hooked on a digital product, but there are also many warning stories about it being misused. Part of using behavior change design to build products is being clear very early in the process about what success looks like and what it does not. Otherwise, you risk participating in an arms race for "most time on screen."

Some Core Tenets

The behavior change design approach in this book weaves together multiple different behavior change theories. Buckle up: This is your whirlwind tour. If you've taken psychology courses, you will recognize several old friends: self-efficacy, social learning, mindsets, and so on. The theory that most heavily underlies the book's organization is the self-determination theory of motivation, which builds on and extends older theories of behavior in a way that's easy to apply to product design.[2] I describe it in more detail in the section called "All About Motivation."

Aside from specific underlying theories or approaches, there are three important points to remember when designing for behavior change:

- **People are different.** There will rarely be a "one-size-fits-all" solution to any problem. It's important to be clear about who you are designing for and what they need. Research is essential to paint the picture of who your users are; because people are different, it's very likely your assumptions based on your own experience or the people close to you won't be true for others.

- **Context matters.** Nothing happens in a vacuum. People's reactions to your product and their ability to take action depends on their situation. Understanding the environment in which people use your product, as well as the circumstances in which they will work on behavior change, should inform how you design.

- **Things change.** The whole point of behavior change is progress. As time goes on, people's needs and situations will probably evolve. The way your product works for them may evolve, too—to the point where your users may even "graduate" to not needing it anymore. Be open to the idea that your users' needs will change over time, and ready to adapt.

The research activities built into the behavior change design process help keep these three points at the forefront of your work.

Terminology

When you're doing behavior change design, there are some words that need to be used in very specific ways. While I recognize that

---

2 Its 40+-year-old evidence base doesn't hurt either.

glossaries do not typically ignite readers' loins, it's really important to know what these terms mean because I use them about six thousand times each in the rest of the book.

The *target users* are the people for whom your product is designed. The target user group should be defined as clearly as possible; this will help with clarity in your research, design, and marketing. For example, Runkeeper's target users are busy people who are interested in personalized routines that fit their schedules and fitness levels. That's a different target user group than Couch to 5k, which is designed for people to ease into running by incrementally building their fitness. The two products both support the behavior of running, but do so quite differently because they're designed for different users. Some teams represent their target users with personas, which encapsulate key characteristics within a fictional user profile. Personas can provide a kind of shorthand to keep target users top of mind during the design process.

For the most part, a *behavior* is something people do.[3] Behavior change, as the phrase implies, focuses on behaviors. Being precise about what behaviors you're designing to change is important because it's very easy to get distracted by related nonbehaviors. In particular, designers may consider *emotions* or *cognitions* as targets to change, like:

- Increasing someone's confidence to do something
- Persuading someone to have a new belief
- Making people feel happier

These are worthy goals. But they are not behaviors.

Why the focus on behaviors over emotions and cognitions? After all, what people think and feel has an effect on what they do. But it's what people do that ultimately affects meaningful outcomes. To take a pragmatic standpoint, the outcomes are what people pay for when they hire behavior change designers, not the feel-good intermediary steps where users gain confidence and embrace positive beliefs. Behavior change design is successful when it saves money or improves symptoms of a disease or makes a process more efficient, and doing those things requires moving the needle on behaviors.

---

3 Some more complex behavior change projects might focus on thoughts or feelings as "covert behaviors"; biofeedback programs are an example. For the majority of behavior change design projects, the more simple definition of behaviors as observable actions helps minimize confusion.

If you're not certain whether something is a behavior or not, a good question to ask is "Can I see this?" Behaviors are observable. Emotions and cognitions are not.

It's entirely possible that a behavior change intervention might target emotions or cognitions as a way to influence behavior. Take confidence; usually, people have to feel some confidence to try a new behavior for the first time. If it goes well, their confidence increases and they're more likely to try the behavior a second time. An intervention that boosts people's confidence to try a behavior might be an excellent way to make that behavior happen more frequently. But the designer has to have the behavior as a target in order to determine that confidence is the right lever to get results.

Every behavior change intervention has one or more *target behaviors*. A target behavior is a specific behavior that the designers are trying to affect—whether they want people to do it more, less, or differently than they're already doing it. One of the very first steps in designing a behavior change product is figuring out what the target behaviors will be. Almost every other design decision cascades from that one, from the data that you'll collect to the features you'll include.

Some behavior change projects focus on getting people to do a behavior a limited number of times. Paying parking tickets is an example; even if someone gets a lot of parking tickets, paying them is a one-shot behavior that doesn't really require any sort of ongoing attention or effort. More often, behavior change is a more complicated endeavor that requires people to make consistent and sustained changes to their lives. Something like managing a complex health condition or socking away enough money for retirement may be a lifelong effort. And those sorts of ongoing changes require people to have motivation.

Motivation can be defined as *desire with velocity*. The way most of us use the word in our daily lives isn't quite right; people might say something like "I'm motivated to be rich," but they aren't really. They *want* to be rich. That's just desire. Motivation, like behavior change design, takes a target behavior as an object. You could be motivated to save extra money each month, find a good fund to invest in, or take a new job with a better paycheck. That's desire with velocity.

If there is such a thing as "one simple trick to changing behavior… forever!" it's connecting people with their motivation. Fortunately, psychology offers designers a toolkit to do just that.

All About Motivation

Psychology offers a multitude of theories to understand what motivation is and how it works. As I noted earlier, my favorite of the bunch is the self-determination theory, or SDT. SDT builds on classic theories of motivation like Maslow's hierarchy of needs, and it plays nicely with concepts like self-efficacy and habit formation. It's also got one of the richest bodies of evidence in psychology, with over 40 years' worth of studies that cover health, education, finance, sport, and a bunch of other behavioral areas. And it resonates with people's lived experiences. When I talk to people about SDT, I can see them recognize the concepts in their own lives. All of these factors make it an excellent starting point for designing engaging digital experiences.

Motivational Quality

In addition to defining motivation differently from the way people use the term in normal conversation, SDT also quantifies it differently. It's typical for people to talk about motivation as something that has an amount. The more motivation someone has, the more likely they are to do something. The self-determination theory of motivation takes that a step further to consider *motivational quality*. It's not just about how much motivation someone has, but also what fuels it.

There are six types of motivation according to SDT that can be arranged along a continuum from *controlled* to *autonomous* (see Figure 1.1). In order from most controlled to most autonomous, the motivation types are:

- Amotivated
- External
- Introjected
- Identified
- Integrated
- Intrinsic

Simply put, the more controlled a type of motivation is, the more it is imposed onto someone from an external source. The more autonomous it is, the more it's generated by the person from within.

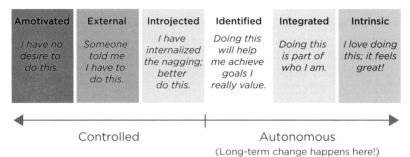

Amotivated	External	Introjected	Identified	Integrated	Intrinsic
I have no desire to do this.	*Someone told me I have to do this.*	*I have internalized the nagging; better do this.*	*Doing this will help me achieve goals I really value.*	*Doing this is part of who I am.*	*I love doing this; it feels great!*

Controlled ←———————————————→ Autonomous
(Long-term change happens here!)

FIGURE DESIGN BY AIDAN HUDSON-LAPORE.

FIGURE 1.1
The controlled forms of motivation are the weakest, but as people find personally meaningful reasons to do something, their motivation takes a stronger autonomous form.

The first type of motivation is really not motivation at all. *Amotivation* refers to when a person has no motivation. People who fall into this category are highly unlikely to take even a first step toward behavior change. As such, it's very unlikely you'll find them registering to use your program.

Sometimes people may be pushed into the behavior change arena by forces outside of their control. Most commonly, these forces take the form of someone nagging or a financial incentive to make a change. Someone who is motivated to try a behavior for purely extrinsic reasons can be said to have *external motivation*.

People may internalize others' expectations of them over time. The word "should" is a clue that this may be happening: "I should lose weight," "I should study for the test." When the external forces causing motivation are internalized, people experience *introjected motivation*.

But what if a person has their own reasons for wanting to do a behavior? The first type of motivation on the autonomous side of the scale is *identified motivation*, when a person sees a behavior as instrumental in achieving something they value. They may not be very interested in the behavior for its own sake, but see it as a stepping-stone somewhere else.

It's possible for a person to see a behavior as reinforcing an important value or part of their identity. Maybe they consider themselves as a kind person, and spending two hours every week at a volunteer project helps them live into that value. When a behavior is important because it supports someone's identity or values, the person has *integrated motivation*.

The final and most autonomous form of motivation is *intrinsic motivation*. This happens when a behavior is enjoyable purely for its own sake. This is rarely seen in behavior change projects; although some target behaviors can be pleasurable, it usually takes some time and training before that's the case. Consider exercise, which can feel wonderful for someone who has a comfortable routine; it's hell for many newbies. On the other hand, many of the "bad" habits that interventions try to break may be very enjoyable for people, and therefore hard to abandon.

More autonomous forms of motivation are better than controlled ones for long-lasting behavior change. Behavior change tends to be hard for people; if their reasons for trying to change are personal and deeply held, they're more likely to help them weather the difficult times. Controlled forms of motivation are more vulnerable to obstacles. Therefore, a goal of behavior change design is to coax people along the motivational continuum away from controlled forms of motivation toward more autonomous alternatives. This is done by designing experiences that fulfill people's basic psychological needs.

Basic Psychological Needs

Yes, people are different from each other, but in some fundamental ways, they're all the same. One of those fundamental ways is that all people share three universal basic psychological needs. The more these needs are supported by an experience, the more people want to engage in it. Basically, support for these needs is what makes an experience fun, interesting, or fulfilling. Because people are hardwired to satisfy basic psychological needs, they are extremely sensitive to cues in the environment that either support or thwart them.

The first basic psychological need is *autonomy*. Autonomy means having control and being able to make meaningful choices. Meaningful choices include which goals to pursue and, broadly, which methods to use to pursue them. Programs that dictate user goals are likely to feel less autonomy-supportive; similarly, programs that give users lots of choices that aren't meaningful, like the color of a dashboard, won't fill this need.

The second basic psychological need is *competence*. Competence is supported when people can see that they're learning or growing with time and activity. People thrive on progress. Regular, clear feedback helps people see what they've done. And designers can help ensure that people make progress by identifying the obstacles stopping them from success and creating ways for them to overcome those obstacles.

The third basic psychological need is *relatedness*. Relatedness is satisfied when people feel part of something larger than themselves. Often, relatedness comes from one-on-one or small group relationships, but people can also get their relatedness fix from being part of a community, feeling connected to a higher power, or having an emotional bond with a pet. People are very good at creating connections, even with inanimate objects, so it's possible to help fill their relatedness tank through technology alone.[4]

NOTE TEACHABLE MOMENTS

Part of the design process involves understanding the circumstances under which someone might use your product. Depending on what your product is intended to do, you might identify opportunities where users are more receptive to the idea of a specific behavior change. An example of a teachable moment is the aftermath of a health crisis; someone who's just had a heart attack may be ready to consider an exercise program that seemed unnecessary before the diagnosis. On the more positive side, someone who's just gotten a promotion and raise at work may be open to assistance in paying down student loans more aggressively. Understanding these teachable moments can inform your marketing and onboarding strategies, as well as the way you structure goal-setting within the product.

Cross-Cultural Relevance

Self-determination theory has been used in research around the world. For the most part, the theory holds up across national and ethnic cultures, as well as across socioeconomic classes. There are some cultural differences in *how* people seek support for their basic

---

4 This is *not* a suggestion that people partner with robots. That's a dystopian future I want no part of.

psychological needs, but those needs themselves are found universally. The take-home message is that SDT is safe to use as a lens for your product development, but be sure to research the specifics with the target users in the target context.

The Behavior Change Design Process

The phases of a behavior change design project look a lot like the phases of any project. At Mad*Pow, where I work, the behavior change design team breaks the process into four phases:

- **Diagnosis** is a research and discovery phase to better understand the problem space, the target users, and the context in which they live and perform the target behaviors. This phase could include primary research like interviewing or observing target users, as well as secondary research, such as literature reviews or reviewing insights from previous projects.

- **Prescription** is a generative phase where potential solutions are explored and product requirements created. Literature reviews can come in handy here, too. They offer a fast track to identifying solutions that are likely to be effective, especially when paired with a framework like the Behaviour Change Wheel[5] that links behavior change barriers to interventions.

- **Execution** is the phase where the product is built, whether that takes the form of visual design, coding, service development, and so forth. Behavior change designers may be active contributors to the build, or they may work closely with other team members to ensure an accurate translation of concepts.

- **Evaluation** refers to the measurement of a product's effectiveness. Although it's represented as the end phase of a four-part process, evaluation is ideally an ongoing activity throughout the design process to maximize product success. For example, it's always a good idea to test early prototypes with users if you can, rather than investing in a full development before getting feedback. Once a product is out in the market, measurement provides tools to continue to iterate and improve it.

Regardless of what your team calls these phases, you probably have a similar process that you use. The labels matter less than what you

---

5 You'll learn about the Behaviour Change Wheel in Chapters 5 and 6.

do during each phase. Behavior change designers typically perform a specific set of activities to accomplish phase objectives. Sometimes these are familiar activities that other design professionals also use, like research interviews or prototype sketching. Other times, the activities are more closely tied to behavior change, such as doing a literature review of previously published interventions or creating an outcomes logic map. What's important is that you are doing the groundwork to understand your target users and their needs, coming up with reasonable hypotheses about how to change target behaviors, and testing the efficacy of your designs.

The tactics and activities in this book occur across the entire product development process. You'll learn about ways to research and understand target users; identify product features to help them accomplish goals and overcome barriers; specify requirements so they can be brought to life in product development; and investigate whether and how a product works to support user outcomes.

The Upshot: You Can Do Behavior Change Design

Behavior change design is the application of psychological science to the creation of products or experiences in order to influence specific behaviors in users. Behavior change designers work to incorporate science to the inner workings of an intervention to maximize how effective it is at producing target behaviors, the things they want their users to do. At the same time, psychology can make any digital product more engaging by supporting users' basic psychological needs of autonomy, competence, and relatedness. Behavior change designers follow a process that includes research, translating findings to product features, and measuring outcomes.

In this book, I use a specific psychological theory called the *self-determination theory of motivation* to talk about how to engage users with a digital product. Self-determination theory's main premise is that the strongest types of motivation come from people's deeply held goals and values. Engaging experiences help connect people's behaviors with what really matters to them. Designers can create those connections by supporting users' needs of meaningful choice (autonomy), learning and growing (competence), and belonging to something larger than themselves (relatedness).

With a combined background in epidemiology, biomedical informatics, and public health, Heather Cole-Lewis is one of the most rigorous thinkers I know when it comes to integrating behavior science with product design. Heather specializes in health, but her advice about subject matter expertise holds true across all fields of practice.

What does it take to do behavior change design?

I like this field because there's always an opportunity to learn more. We sit at the intersection of a lot of different worlds. One person might have all of these different areas of training; however, to get the work done at the rate at which it needs to be done, we have to work together. So we can come to the table with our existing knowledge and skill sets acknowledging that there are other pieces of the puzzle that have to come together in order for us to build engaging, successful, cost-effective solutions. You have to be willing to teach *and* to learn from other people at the table that are just as skilled in their particular areas.

What do people misunderstand about behavior change?

We all are humans and we all behave, so a lot of behavior science feels intuitive. But there is a true science when it comes to understanding behaviors, contextualizing them, figuring out what to do about them, and ultimately measuring the effect. That's where the skill set of a behavior scientist comes in. A lot of times people expect behavior science to happen overnight. The challenge is getting them to understand the rigor that it takes to get to a great solution. The results of health behaviors often take some time to see. But a great behavior scientist can incorporate innovative research methods into a design process to help the team move quickly and cost-effectively.

With health behavior, there are a couple of different levels of engagement.[6] The first thing that has to happen is that someone has to use your solution. Next, once you get them to pick it up and use it, the clicking and swiping has to be toward a very particular end—exposure to the techniques that address barriers and facilitators of health behavior. Finally, you have to ask "Did their health behaviors improve or change?" Once people have a better understanding of engagement with the technology versus engagement with the health behavior, they start to have more hope for better outcomes that could happen with engaging technology that has great science built into it.

Also, people have to realize that health behavior science is about choice, not tricking people into doing things they don't want to do.

---

6 Learn more in Heather's paper about engagement at https://formative.jmir.org /2019/4/e14052/

What is your behavior change design process?

Start by building a theory of change or logic model.[7] When you build for behavior change, you start with the challenge that needs to be addressed and then the behaviors associated with it. Go backward until you get to the actual behavior that needs to happen. Then you categorize the world of reasons why a person does or does not do a behavior, determinants of behavior, and see which ones are most important for this particular group of people. Once you've got behavioral determinants mapped out, it helps you understand which behavior change techniques to use.

The logic model is going to help you understand how everything you add to this intervention represents a behavior change technique that goes toward one of those determinants to ultimately change behavior. As long as you're laying out your assumptions in the logic model, then you're well set up for understanding if your solution is going to work or not. In order to build a great intervention, you have to think about how you're going to measure it from the very beginning. Otherwise, you're not going to be able to analyze how it works in the end.

What is the business value of behavior change design?

One reason you build a logic model on the front end is because then you can understand where to spend your budget first. A behavior scientist can help you evaluate and build out your solution in an iterative fashion to understand if you're moving in the right direction before you build out the whole thing. They can also help to make sure that if the product is digital, you build the data framework that will help you evaluate it.

Some behavior scientists are still learning how to articulate behavior science in business terms. That communication should happen with an eye toward value, or whatever the ultimate end point is. If you're working in a commercial space, why does this behavior science matter? Is it because you're getting people to improve their health in the long term? In the short term, is it about building awareness of a health issue and showing that the brand cares about the consumer's experience with that health issue? The science is not always enough. It has to be about the science and the business case.

 *Heather Cole-Lewis, Ph.D., MPH, is Director of Behavior Science at Johnson & Johnson Health and Wellness Solutions. She holds a Ph.D. in epidemiology from Yale University, an M.A. in biomedical informatics from Columbia University, and an MPH from Emory in behavior science and health education. The opinions expressed are Heather's and not endorsed by Johnson & Johnson.*

---

7   See Heather's paper at https://bit.ly/35aAGOu

CHAPTER 2

Pictures of Success

Measurement and Monitoring

How do you know if your product works? You measure! Measurement is critical for any sort of design, and even more so for behavior change design. The process of measurement is front-loaded, which is why I'm putting it at the front of this book. It requires detailed, thoughtful planning early in the product development process. It may feel like it's taking a lot of time—perhaps too much time. But if you do the planning well, then you will avoid all sorts of missteps down the road. You'll make better design decisions early in the process, saving expensive and time-consuming rework later. And you'll save time when you're actually collecting outcomes data; it's a matter of executing to your plan. This chapter focuses on the preparatory work that goes into measuring and monitoring the performance of a digital product.

Behavior change designers must think not just about the typical metrics associated with digital products (like registrations and revenue), but also the long-term outcomes associated with getting people to do something differently over time. A corollary to how long it takes for behavior change to happen is that it also takes a long time—weeks, months, or even years—to measure the most meaningful outcomes and get them into shape to share. But never fear, there are plenty of indicators available earlier in the process that can help you gauge a product's success.

Measurement isn't just for proving the effectiveness of your product. It can also help you figure out how to improve it over time. By monitoring the right data, you can identify opportunities to build new features or content that helps users achieve their behavior change goals. You can also see if there are any clunkers in your product—features that are unloved, unused, or unneeded.

Writing Your Outcomes Story

The foundation of a strong outcomes story is to start at the end.[1] Very early in the process, sit down and write the story you'd like to be able to tell once your product is on the market and thousands of people are using it. This could literally be a story—a helpful approach can be to write an imaginary press release or magazine article about your

---

1 Matt Wallaert's book of the same name (*Start at the End: How to Build Products That Create Change*) covers the process of designing for behavior change in a way that, not surprisingly given his title, also starts with creating an outcomes story. Which makes sense, because it's the right way to do behavior change.

successful product and highlight what it has accomplished. If you're more visual, create a storyboard or journey map that illustrates an ideal user experience with your product over time, including the indicators that the product is working. Invite others on your team, or your clients, to weigh in on the story so that it captures all of the most important success metrics.

At some point, the story can be formalized into an outcomes plan. One way to represent this plan is as an outcomes logic map, a document that shows the types of outcomes your product might produce over time. The outcomes logic map is a tool used in program evaluation research. One reason why it can be so powerful for digital products is that it considers how a product gets implemented and used, not just whether it's effective in isolation. Data like click metrics, bounce rates, and drop-offs don't just show usage; they can also help designers understand how their product is being used and make adjustments that improve outcomes.

Figure 2.1 shows a generic outcomes logic map without the information filled in. The word "logic" refers to the fact that each measurement should logically connect to the others in the sequence. There's a concept in psychology called the *mechanism of action.* That refers to anything that facilitates an outcome happening. For example, if the outcome is lower body weight, the mechanisms of action might include physical exercise and changes to diet. You'll want to identify mechanisms of action for the outcomes in your story. Measuring them will give you a way to test whether your product is doing what you expect it to do, and make adjustments if not. You'll include mechanisms of action on your outcomes logic map as links between people using your product and the results they eventually achieve.

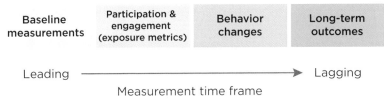

DIAGRAM BY AIDAN HUDSON-LAPORE.

FIGURE 2.1

At a high level, an outcomes logic map describes the data that should be available over the lifespan of a product to understand whether it will achieve its intended results.

This map gives a visual indicator of what types of things to be measuring at different periods of time. If your measurements at these points suggest that the product isn't meeting your goals, it's an opportunity to dig deeper and figure out why not. Plotting outcomes over a timeframe also helps set expectations about when to determine whether or not a product works. It doesn't make sense to say a language learning program is unsuccessful if its users aren't fluent in Greek after three weeks of use, but that same data from people who are three-year veterans might be more meaningful.

Your first attempt at an outcomes logic map will not be perfect. That's okay. The critical thing is to go through the exercise so that your design process is organized and focused. You can and will update your outcomes logic map as you learn about your users and their needs, both during your design process and after your product is in market. But without a basic map in place at the outset, it will be difficult to know what questions to ask and what key performance indicators to build toward.

Here's how to create an outcomes logic map.

Define the Long-Term Success Metrics

Look at the ending you've written in your product success story. What are the most critical metrics your product would achieve if it were successful at changing people's behavior in a sustained way? Think about outcomes that might take months or years to accomplish, ones that would make someone look at your product and say "wow." Write them down.

In behavior change, long-term success metrics generally fall into one of two buckets: domain-related or financial (i.e., return on investment). The domain-related outcomes vary widely, depending on what exactly the product is intended to do. If you were working on a health-related behavior change product, your long-term success metrics might include biometric data or anything that could be measured through medical tests, like someone's weight, blood pressure, or the level of nicotine in someone's blood. There are some health conditions, especially in mental health, that don't have a physical test, so the outcomes there might be something like changes in people's scores on a depression inventory. Health outcomes might also include performance indicators, like how fast someone runs a 5k or whether they're able to complete a memory test successfully after brain training.

What if your intervention isn't about health? If your intervention focuses on financial behaviors, long-term outcomes might include balances in someone's savings account or the percentage of employees who have opened a 401(k) retirement account. For education, long-term outcomes might be achievement levels on a standardized test or the percentage of learners who could successfully complete an applied task. For sustainability, it could be the number of meatless meals eaten per week or the pounds of trash diverted into recycling or composting. You get the picture.

The financial outcomes often include money saved by using the medical system less or better; for example, a successful intervention might have people using the emergency room less often but the pharmacy more, because they're taking their medications as prescribed. Sometimes the financial outcomes include increased revenue because more customers are buying the product as a result of its success. In digital projects, a common success metric is having customers choose to use the digital channel to complete tasks instead of using more expensive resources like call centers. Defining exactly what the financial success metrics for a given product look like require you to know your industry and your client well.

As you write down what long-term success looks like, be as specific as possible. Don't just write "blood pressure"; write "reductions in blood pressure" or "percentage of people whose blood pressure changes from being high to normal." The more specific you can be, the better able you'll be to get the right data to tell your story.

TIP LESS CAN BE MORE

Especially for newer products, a simple outcomes story may be more powerful than a kitchen-sink-style one. Amazon once sold only books; Google used to be just a search engine. Their compelling early successes gave them a platform to expand their purviews. Just like a minimum viable product focuses on the most essential features, your first outcomes logic map should focus on only a handful of truly critical outcomes.

Identify the Behavior Changes Needed

Your long-term outcomes don't happen by magic. They happen by behavior change on the part of your users. In most cases, this behavior change needs to be sustained over time. To take the example of lowering blood pressure, in order to see measurable differences on tests, people might need to eat differently, start walking regularly, and take their medication every day for weeks, months, or longer. It's not a one-shot deal.

Focus the metrics in your outcomes map on behaviors that your product can reasonably be expected to influence. If your product is a medication management app for people with high blood pressure, your outcomes plan should focus a lot more on medication-related behaviors than the exercise and diet pieces. You might want to look at those behaviors, too, just to explore whether your users are also making those changes—sometimes people on a behavior change kick will have a positive spillover effect to other behaviors—but don't hang your hat on them since it's not what your product is designed to do.

You'll want to make a judgment call about how granularly to measure behaviors. For example, if taking medication is an important part of achieving outcomes, you *could* just list "take medication as prescribed" as the behavior of interest. But if you know your users probably don't take medication yet, you might also include "make an appointment with a doctor" and "fill prescription at pharmacy" as additional steps. That will remind you to coach users through those steps within the product and help you identify where the pitfalls are if people don't actually end up taking the medication.

It's very likely that your product can't directly measure the behavior changes that lead to outcomes. That's fine! It's very typical, in fact, since most behavior change happens out in the world and not online inside a digital product. You'll find other ways to measure whether people are doing these behaviors. What's important is identifying the behaviors that need to happen for your success story to come true, so you can design your product to affect them.

Determine How to Measure Exposure

The term "exposure" here is a fancy way to ask if people are actually using your intervention. There's a famous quote from the former Surgeon General C. Everett Koop that I often use when presenting about this topic to healthcare audiences: "Drugs don't work in patients who don't take them." The modern corollary is "Interventions don't work in people who don't use them."

You'll want to measure usage for at least three reasons. First, theoretically, your product won't do anything if people aren't using it. Second, you need to be able to show that people used your product as part of their behavior change process to tell a compelling story about its success. If you put a product out in the world, and a year from now your long-term success metrics have come true, but the people showing those changes never used your product, you'll have a hard time convincing anyone that you had anything to do with the changes. Third, usage is a "leading indicator"; you can measure it almost immediately after putting your product out into the world. Leading indicators are your earliest evidence of whether a product is successful or not.

Some of the common measurements you might include as leading indicators include:

- App downloads or installations
- User accounts created
- Logins or sessions started
- Actions taken within the program (e.g., articles read, videos watched, action steps checked off a list)
- Return visits

NOTE DON'T STOP HERE

A common pitfall product teams make is focusing too much on leading metrics.[2] There are all sorts of reasons why this happens; they're relatively easy to measure, and they enable quick reports back to leadership if they want to track how a product is doing. While sometimes people struggle to understand the complexity of the more meaningful behavior change outcomes, most people quickly get the importance of the number of users or the frequency of use. It's okay to use your leading metrics as a success indicator, especially early in your product's life, but don't lose sight of the more critical lagging metrics. They tell a *much* more compelling story.

---

2 Laura Klein has an excellent overview of some of the problems people run into measuring leading metrics in her book *Build Better Products*, including the issues that can be caused by trying to game the leading metrics without paying attention to the lagging metrics they predict.

Some of these metrics may be familiar to you from another metrics planning tool, the conversion funnel (Figure 2.2). If your team uses a conversion funnel to track marketing and acquisition, you can incorporate it into a larger outcomes logic map. Your funnel won't be so funnel-shaped anymore, but it will do the job of helping you track the right metrics across the product lifecycle.

Sees marketing & becomes aware

Visits site

DIAGRAM BY AIDAN HUDSON-LAPORE

FIGURE 2.2

A conversion funnel is a type of metrics planning tool that helps teams gauge how well they are reaching, acquiring, and retaining product users.

Signs up for product

Repeats use

Keep in mind, too, that *more* does not necessarily mean *better* with exposure metrics. Most behavior change interventions need a certain amount of time to work, and more is overkill. If you know what the right "dose" of your intervention is for your users to accomplish their goals, make that the target and not a session more. And if you don't know, make an educated guess, do some testing, and iterate over time.

Fill in the Specific Data Needed

The last step in completing the outcomes logic map is to fill in the specific data you'll need to collect to determine if you've achieved each outcome. "Specific" is key here: "blood pressure readings" is not as good as "self-reported blood pressure readings using a kiosk at the pharmacy."

In the process of listing out the data you need, you'll be able to identify information you can collect through your product itself and the information that you'll need to get outside of your product. Anything that happens inside the product needs to be accounted for in the design requirements; for example, if you need to have users answer certain questions every two weeks, you'll need to make sure those questions are included in the product, along with a scheduled prompt so that people answer them at the right times. Anything that happens outside of the product will need to be accounted for in other ways. I'll cover some of those in the section on "Evaluating for Effectiveness" later in this chapter.

Get a Baseline

Outcomes stories are really stories about change. There's "in the beginning," the bad old days, and then the happy ending of today's outcomes. In order to tell the change story well, you need to know what "in the beginning" looks like. That's why, once you know what metrics you'll be collecting for outcomes, you should measure them at the outset to establish your baseline.

If you've ever used an app or a program that starts with a long questionnaire, one of the things it's doing is creating this baseline about its users. If you have information about your specific users before they became your users (or shortly after), you can do what's called a "within subjects" analysis where you look at changes *in the same people* over time. If you don't, you can assess a similar group of people who aren't your users and compare them to the people who are your users. This is called a "between subjects" analysis. Later in this chapter ("Evaluating for Effectiveness"), you'll see how both types of analysis can be used to tell part of an outcomes story.

Plan Your Analyses

You should have a plan for data analysis before you collect any data. Having an analysis plan up front helps you collect the right data to tell your stories, in a large enough quantity to yield meaningful results, and in a format that is useful and usable for you.

Here's an example: Let's say you are doing that blood pressure management app I've used as an example already in this chapter. You want your users to take their medication more often so they ultimately get their blood pressure numbers into the normal range. So, of course, you will measure how often they take their medication. Now, you know that people are likely to slip up with behavior change, especially when they're first starting out. So you want your measurement to be sensitive enough to show small improvements, even if your users still aren't taking their medication perfectly. Knowing that, how should you measure their medication behavior?

You could have them count the pills left in their bottle at the end of the month and then subtract that number from the original quantity. That would get you a number for each user representing the number of days they took their pills. Or you could ask people at the end of the month to answer one question: "Did you take all of your medication as prescribed this month? Yes or no." That would get you a yes or no for each user.

Since you know you want to show small improvements and not just total success, the first option is much better. It will let you show changes to the average number of days your users took their medication in a month. A change from, say, 21 days to 23 days will be detectable with your data. If you'd asked the yes/no question, all of the people who have improved their medication behavior but not perfected it would look like failures, when, in fact, they are making positive progress. Having the more granular data will also let you look more closely at the people making progress to understand what's working for them and where you might be able to improve the product to help them more.

The way you collect data has a huge influence on what type of outcomes story you can tell. Writing the story first and understanding what analyses need to happen for you to tell it will help you ask the right outcomes questions.

Chances are, you aren't a stats whiz, and you'll have some questions about how to do your analyses. Depending on the complexity of the project and the type of data you'll collect, you may want to work with a colleague who has strong quantitative skills or even a data science professional. For simpler projects, you can find excellent resources on how to analyze different types of data for different objectives online.

Once you've specified your data, you have a complete outcomes logic map for your project. Figure 2.3 shows what that might look like for the hypothetical high blood pressure product that focuses on helping people take their medication more regularly.

Socialize this document with the other members of your team, especially anyone working on building or marketing the product. Don't be afraid to tweak it as your product grows and you learn more about your users. As your product makes its way into the world, you can start associating actual data with the outcomes on your map. If they don't support the story you want to tell, then you know it's time to look at your product and make some changes. But first, let's talk about where you can get the data to tell the story.

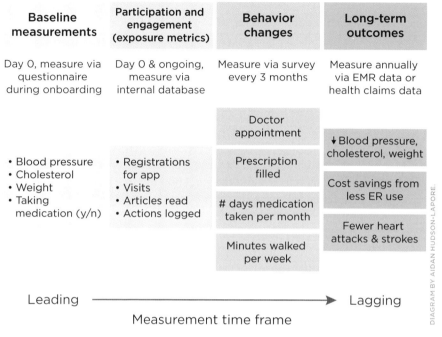

FIGURE 2.3
A completed outcomes logic map that includes specific data and measurements for baseline, exposure, behavior changes, and long-term outcomes. For more complex products, the map will also be more complex.

NOTE CHANGE IS NOT FAILURE

It can be emotionally hard to get data that suggests something you designed wasn't quite right for users. But informed changes to a product are a healthy part of the design process. It's much better to learn early on that a feature doesn't help people, so you can stop maintaining and iterating on it, than to continue to invest time, money, and resources on a dud. And early research can inspire new features that make the product. Did you know the first iPhone didn't connect to an app store?

Evaluating for Effectiveness

The most important measure of whether a product is effective is that it produced the changes it was supposed to in users' behavior. Effectiveness is a fancy way of saying, did the product work? Behavior change interventions differ from other sorts of digital products in that their ultimate intended effects take place off the screen in almost all cases. Constraining your measurement practice to the digital product itself means that you won't be able to detect the real-world outcomes that are hopefully part of your results.

Fortunately, there are several ways to collect data outside of the product itself that will help you understand the product's effects. This section doesn't cover them exhaustively, but I've highlighted a few of the most common and effective methods of figuring out whether and how a product works. I've also focused on research methods that are less common in UX and design, knowing that many in-depth resources exist for the more typical research toolkit. These methods do require an investment of time and resources above and beyond actual product development and customer acquisition costs, but they offer a huge rate of return if your product is able to attract investors or users with their results.

It also helps not to think of effectiveness research as a one-shot deal. Over the life of your product, you'll do many studies that each tell a piece of the story. In the interview at the end of this chapter, Cynthia Castro Sweet from Omada Health talks about how to layer research over time; it's a good reminder not to feel overwhelmed by the idea of designing the perfect study.

The Gold Standard: Randomized Control Trials

Randomized control trials (RCTs) are the gold standard of outcomes testing. Realistically, most companies won't have the resources to do RCTs on their product, but understanding how they work can help with planning more realistic, scaled-down studies that target similar sorts of understanding. RCTs are in a sense the most scientific way to see the effects of an intervention on people's behaviors. RCTs are a way to look at both within-subjects and between-subjects changes by looking at changes in people who use a product, as well as comparing them to people who do not. They include three elements, handily summarized in their name: randomization of research participants, a control condition, and a defined trial period.

Randomization means that the people who participate in the study are randomly assigned to either use the intervention or use something else. The randomization is important because, with a large enough sample size in the study, it theoretically eliminates the explanation that it's something about the person that leads to the results. In the real world, people often self-select into experiences or products. They pick the things they enjoy and avoid the things they don't. But then if something really works for them, you don't know for sure that it's because the product is good. It might be that this person liked a crappy program enough to stick with it long enough to get results. So in an RCT, people are randomly assigned to the product they use.

Control refers to eliminating other explanations for any outcomes. People in the *control condition* do not use the product, but are assigned to use something else that is reasonably similar. Research has shown that just becoming more aware of a behavior can change it. To rule out the explanation that an intervention works only because people are paying attention, the control condition will also do something that increases their attention. In a sleep-coaching program I worked on, the control group in the RCT completed sleep diaries and questionnaires, while the experimental group used our program. Because both groups were spending roughly the same amount of time each week thinking about and logging their sleep, it was easier to say that improvements were because of our program and not because of the heightened attention to their sleep habits.

Finally, *trial* just means that the study takes place over a contained period of time. Usually, the researchers who run an RCT will recruit people to participate for a defined length of time, chosen because it's long enough for the outcomes of interest to be detected. At the start of the study, the researchers will take baseline measurements from all participants, regardless of whether they're in the experimental or control condition. Over the course of the study, additional data might come from the product (or the control equivalent), as well as from additional surveys or observations from the researchers. Finally, at the conclusion, the researchers redo the measurements from the baseline in order to see what changes have taken place.

TIP WORK WITH THE PROS

Designing, running, and analyzing a real RCT is a specialized skill that you don't find in most product organizations. If you want this kind of research, consider hiring a contract research organization (CRO) or partnering with researchers from a university.

RCTs are the closest thing in the behavior change world to prove that something works. They can be expensive and time-consuming, but when they're successful, they can set a product apart from its competition.

Real-World Comparisons

Real-world comparisons use an existing comparison group instead of a randomized comparison of people using a product to people using something else. Unlike an RCT, you don't control what programs, if any, the comparison group uses. You do try to establish similarities between the comparison and test groups; for example, you might compare people using your new walking program to people who have a Fitbit. Often, a real-world comparison is much more feasible for companies to do than an RCT. They provide strong evidence for a product as long as the research team takes steps to rule out alternative explanations for results.

Here's an example from my own work. The members of a large health plan had access to digital health-coaching programs to target a handful of chronic conditions like high blood pressure and diabetes. Some of the members used the programs; others did not. The health plan wanted to figure out if the programs made any difference. So they were able to identify a couple thousand program users in their database and separate them out. Then they identified an equal number of nonusers in their database who were pretty similar to the users at the outset; a similar mix of men and women, average age, and level of disease symptoms. Finally, they compared the data for the two groups at different time points to see if their outcomes progressed the same way or if one group looked different from the other. They found that people who used the programs ended up having less severe disease symptoms (and fewer healthcare expenses) in the years after the baseline data was collected. Because the comparison group was selected to be as similar as possible to the user group, the health plan had confidence that using the programs made a difference in people's outcomes.

> **TIP** USE A THIRD PARTY
>
> In the example of the health plan study, my company made use of third-party data supplied by our client: their claims database. Third parties can be an excellent source of outcomes data, depending on what your product does. If you don't have access to

You don't need to have such robust data to do a real-world comparison, although obviously richer data means a better story. You can still make a case for your product using publicly available information. How do your users look different from nonusers who are pretty similar to them in terms of the behavior being changed?

Case Studies

Case studies are specific stories about the use of a product. They can follow an individual user, or a group of users, as they interact with a product. Case studies don't necessarily prove whether or not a product works, but they can provide an engaging glimpse into what the product might provide to users. The data in a case study often takes a before-and-after format looking at the same group of people over time that shows how the variables of interest changed after a product was introduced.

Although their scientific value is limited, case studies have some advantages. They offer teams an opportunity to highlight their products in a specific, contextualized way that may help potential customers envision themselves as users. They can also demonstrate aspects of the product, like how it gets implemented and rolled out, that might not be examined in an RCT or other purely outcomes-focused study. And, for many of the same reasons that normative feedback and social comparisons are compelling for users, case studies can be extremely effective in building your business. Clients want the shiny new toy they see in a well-done case study.

Surveys

As a user of technology, you've probably received invitations to complete surveys related to a product you're using. These surveys are a way for companies to learn something about their users that isn't detectable within the product itself. They can be a helpful tool because they're inexpensive to develop and distribute, so you can potentially reach a large number of users quickly and cheaply. Surveys are also an opportunity to ask people about their behaviors outside of your product that may influence their results. Do resist the temptation to ask lots of questions that you don't have a clear reason for wanting the answer to; overloading the survey makes it more likely people will not complete it, and you'll be stuck with a bloated data set that's not very useful.

The disadvantages of surveys include that response rates tend to be low (less than 20% on average), and you're relying on respondents to report their data honestly and accurately. Even people with good intentions may misremember details about their behaviors,[3] while others will be motivated to make themselves look good with their answers. Because of the potential for accuracy and honesty issues, one of the best uses of surveys is to assess user satisfaction. People usually have good insight into how much they enjoy a product, and they're often willing to share.

Surveys are often incorporated into other research approaches such as RCTs. However, they can be used as a stand-alone research method. They're a flexible tool that can be useful both for uncovering product efficacy and learning how to improve a product, which is covered in the next section.

Evaluating for Iteration

Measurement is not just for outcomes. It can also help you determine how to evolve your product over its lifetime. If your users aren't achieving the results you'd hoped for, or they're not doing the behaviors you expected them to do, it's an opportunity to dig deeper and see how your product might be better able to meet their needs.

When your product is off track from the outcomes logic map you developed, it's an opportunity for further research. You can run a survey to see what users think about their experience, or reach out to a handful of users for individual conversations. A focused

---

3 Ecological momentary analysis or diary study tools are designed to overcome this barrier. These methods alert people to provide data in the moment about their behaviors, thoughts, or feelings. These methods can provide excellent insight when context really matters, but may be expensive or complicated to use.

review of your product from either internal team members or an external subject matter expert could also help identify areas that need improvement in order to achieve outcomes. To go back to the blood pressure example I've been using, if you tell people to take medication but don't give them any information about how to *get* the medication, they may not be able to follow through. Identify the roadblocks that aren't being overcome with your product and see what you can do about it.

It's often a good idea to deliberately build test and learn cycles into your product development process. With digital products, the first release is rarely the full product vision; digital development lends itself to iterations. Whether you position the first release as a pilot or a milestone on a progressive roadmap, take the attitude that you will learn things from your users that lead to changes in the product.

There are many types of measurement that fit under the "evaluating for iteration" heading, but are outside the scope of this book to review in detail. These include desirability or usefulness research, which investigates whether a product is appealing to users and helps address their needs; usability research, which focuses on whether people can accomplish tasks within the product; and A/B testing to see which versions of a design are more effective with users. Whatever research methods are in your toolkit, they can probably be used with an outcomes logic map to continually improve your product.

Mind Your Research P's and Q's

How do you make sure that your research is done well and ethically? Generally speaking, academic research and product research follow two different but parallel lanes on the highway of review and oversight processes. You'll want to be clear which of those lanes to occupy with the research you're doing as part of your product development process. The checkpoints you'll hit along the way will help ensure that your research is done ethically and correctly, and will reduce the chances that you lose users' trust through a misstep.

I'm using the phrase "academic research" to describe any research done for the purpose of increasing scientific knowledge without direct product implications. Often, this is the type of investigation done by people at universities or research institutions, but sometimes companies will do it, too. Academic research often makes its way into the world through peer-reviewed journal publications that are theoretically available for anyone to read.[4] Your outcomes research may fall into this category; your product iteration research probably won't.

TIP ASK AND YE SHALL RECEIVE

If you're interested in an academic research paper but can't find the full text online, reach out to the author(s) directly. Twitter is a great way to do this, or you may be able to find their email addresses on the website of the organizations where they work. Most researchers are happy to share their papers if asked.

Importantly, when a team starts the process of planning an academic-type study, they'll have a group called an Institutional Review Board (IRB) look over their protocols and materials. Most research institutions, including universities, have their own IRBs that are free for affiliates to submit to. There are also independent IRBs, which charge small fees to review study proposals. The IRB's purpose is to make sure that any people who participate in the study are treated ethically. IRBs pay attention to details like whether people are compensated fairly for their time in the study, whether participants receive the information they need to understand what's being asked of them and make an informed decision to take part or not, and whether they get the information they need to ask questions later if they want to.

Here's an example of something I've been asked to fix as part of an IRB review for a study. I was giving people feedback on a puzzle task they'd just finished. Half of the participants were told they'd kicked the puzzle's ass, while the other half were told they'd just proven themselves to be the world's worst puzzlers. The feedback had no relationship to their actual performance. The IRB pointed out that the people getting the negative feedback might be in a bad mood afterward. They asked me to include something in the study to make them feel better. Their proposed solution? After getting the fake feedback, everyone watched a video of puppies and kittens. They were confused by the abrupt segue, but delighted by the cuteness.

---

4 I say "theoretically" because accessing peer-reviewed journals can be tricky. They are usually not free to read, and an individual article can run upward of $30 to purchase from the publisher. University libraries often hold subscriptions, but if you don't have an affiliation with the school, good luck accessing their database.

Product research, in contrast to academic research, is done specifically in order to improve a product or service. It is unusual for this type of research to be published someplace where a general audience could read it. Its audience is usually product teams or other organizational decision-makers who will use the information to make decisions about a product feature, roadmap, or investment.

Usually, there is no IRB oversight of a product research study. There may be an internal team who reviews the proposed study protocol to make sure that it meets the needs of the organization and follows ethical processes, but that doesn't always happen.[5] If there's not a formal review process set up, it's still a best practice for an internal team to consider potential pitfalls. Specifically, the team should consider whether users who are part of the research will be put at risk by participating. How will their privacy be handled? Will they experience anything that might upset them or make them feel taken advantage of?

If a product team decides to do research that they might also try to publish in a journal, the best practice is for them to go through both the academic and the product research paths prior to launching the study. Their internal teams will still do their review of the research protocol to make sure that it supports their goals, but an IRB will do an additional review with an eye toward ethical issues.

So, if you're planning research that will add to the general body of knowledge beyond your specific product, have an IRB review your protocol before you begin collecting data. If your research is purely for product development purposes, you probably don't need an IRB. If you're thinking about publishing the results of your study in a journal, that's a strong signal that you should be talking to an IRB.

The Upshot: Metrics Tell Your Story

Having a measurement plan is crucial for the success of your product. You want to tell a compelling story about why your product is great, and the data you gather with your measurement plan will help you tell that story with conviction. Metrics allow you to determine whether your product works, how much people like it, and what the

---

5 In my experience, the companies most likely to have a robust internal review process are large and in high-risk industries, such as pharmaceutical or financial services companies.

most effective ways to improve it would be. If you have a B2B model for your product, metrics help you sell yourself to companies who want to get the positive outcomes you can provide for their people. And whether you're distributing through a B2B or B2C model, success stories interest people in becoming your users.

Perhaps counterintuitively, the most effective metrics are planned at the very start of product development. Doing this ensures that you can build the right hooks into your product to collect the data you'll need, and that you include the right content and features to achieve the results you want. A tool like an outcomes logic map can help you plot out all of the steps that will need to happen to make your product effective. It will guide you in effectiveness research to determine how your product works, as well as research to investigate what your next iteration should look like. And the upfront planning will help you work effectively with IRBs or other reviewing bodies to ensure that all of your research is done in a way that respects users and maintains their trust.

In the digital health world, one company stands apart for the strength of its outcomes story: Omada Health. As of this writing, they've published 11 studies in peer-reviewed journals (that's a lot), and have launched a large randomized controlled trial of their diabetes prevention product. Their work on creating an outcomes story has translated to business success, with Omada boasting one of the highest venture capital fundraising totals in digital health and a top-notch list of clients. A driving force behind Omada's ongoing research program is Dr. Cynthia Castro Sweet. I was interested in talking to Cynthia to learn how Omada has become a leader in telling their story with data. What can other teams learn from Cynthia's experience at Omada?

How can you leverage existing evidence?

There's a big body of literature out there around diabetes prevention programs, but we need to show what our specific product does. You can draw a line from the original Diabetes Prevention Program (DPP) format to the way Omada has implemented it and show apples to apples. By producing our own evidence, we are instilling reassurance that we've held up the integrity and are faithful to the essential elements that made that product or service work in an older, traditional format.

Well-designed, well-conducted scientific study of the product is necessary. You may be able to break into the market with a really cool product, but you're not going to make bigger transformations unless you follow some of the traditional rules of health care. If you're taking something that came from a more traditional format and you're bringing a technological revolution to that, you still need to prove yourself and show what your product can do.

How can you balance research with product development?

I use the phrase "pragmatic scientific rigor" pretty frequently. What I aim for is the best science possible, understanding all the other conditions that are in place for whatever stage the product is in. Sometimes, it's necessary to get your product out there while there's interest in it, even if you haven't done years of highly crafted validation studies on it. A lot of behavior scientists struggle with dialing back their desire to be pristine with their science and trying to marry good science with the other pressing contingencies of product development.

It's part of my job to understand our product vision, what we're aiming for, what the benchmark is in traditional medicine that we are targeting with our program, and when those elements are scheduled to launch. Then I know *what* I can make happen and *when*, in terms of evidence generation.

How do you use user data for research?

Our standard setup allows us to use participant data for our product improvements. So we're watching to see how they're responding to different features and what's being used and what isn't to help us improve the product. When we want to use their data for public-facing research or evidence, then we need to layer in a separate level of consent, permissions, data use, and sharing. I need to show that I've done my due diligence and gotten permissions, and am acting responsibly and ethically in the use of that data. I use an IRB for using my own company's data quite often. We need to be very mindful about when we've collected permission to use data and for what.

How does the science message change for different audiences?

We actually write the same story for different audiences. It's more of a process of distilling and tailoring the right message for the audience. We tend to come out with what we call our power statements, undisputed facts. Then our communications, marketing, brand, and creative teams will craft them into messages for different audiences. Then we have another team of people who review from all angles to make sure what we're saying is true, accurate, appropriate, and can stand up under scrutiny.

And it's not always just about the evidence. There's the user experience, the implementation process, the marketing—there's a bunch of other pieces that you have to weave together to tell the right story. We'll shape the message toward what is most salient and what information we think is most accessible and important for the audience and their decision-making.

Cynthia Castro Sweet, Ph.D., is a health psychologist and behavior scientist. After earning her Ph.D. in clinical psychology from the University of California, San Diego School of Medicine, she took a series of research roles focusing on improving health habits in diverse populations. Cynthia worked for the Stanford Prevention Research Center before joining Omada in 2015. She is now Omada's Senior Director of Clinical Research and Policy. Cynthia's research at Omada focuses on externally validating the efficacy of their programs, particularly their flagship CDC-recognized digital Diabetes Prevention Program.

It's My Life

Making Meaningful Choices

A key ingredient in successful behavior change is motivation. What's really important about motivation is not how much of it people have, but its quality. The people who will be most effective at making a behavior change and maintaining it over time will have one of the types of autonomous motivation, where their reasons for the behavior are tied to their values, goals, or identity.

As behavior change designers, you can create products that bring people's values, goals, and identities to the surface so that it's easier to connect their behaviors to them. You can also offer people opportunities to make choices about how they pursue behavior change. The more people have chosen a path as a result of their own free will, the more resilient their commitment will be. Someone who has knowingly chosen something will be more willing to stick with it if it becomes difficult. Finally, being clear with your users from the outset about what your behavior change program will involve helps them make informed choices and ensure that their experiences and expectations are in alignment.

Ownership Is Key

The ability to make meaningful choices is one of the most important motivational factors people can have—it supports the basic psychological need of autonomy. Meaningful choices start from the most important goals, like choosing to try to change behavior in the first place, and trickle down to supporting goals, like how to go about making change happen. As a general rule, people are more likely to stick with choices they've made themselves than with ones that have been imposed on them, especially when something is challenging.

The reason why? If someone's working toward a behavior change goal and they hit an obstacle, they'll need a good reason to stick it out. A personally meaningful reason for the goal will provide that. "My weight loss app told me I have to do this" will not fortify people to overcome tough temptations.

You can get people started on behavior change by imposing a goal on them. It happens all the time. How many people embark on a diet or an exercise regimen as a New Year's resolution because that's what they're supposed to do? But if you want people to stick with the change, they eventually have to find their own reasons to do so.

Letting people make their own choices about what goals to tackle is a first step in long-lasting behavior change. In the example from Pacifica, a stress reduction app in Figure 3.1, new users are asked to articulate what they hope to achieve from using the program. EasyQuit, an app for smoking cessation, keeps it even simpler and invites users to generate their own list of motivations for quitting cigarettes (see Figure 3.2). You may not need anything more complicated than these types of question to get your users thinking about what their goals are.

FIGURE 3.1
Pacifica's users are asked to describe their goal for using the program as part of the onboarding process.

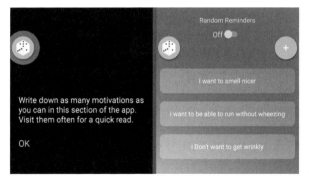

FIGURE 3.2
EasyQuit asks users to generate their own list of motivations for behavior change, and then it offers the option of having those reasons randomly sent as reminders.

Not all choices are meaningful. While users may appreciate being able to choose the color of their program dashboard or the avatar they use, these sorts of aesthetic options don't typically tap into any deeper meaning. In terms of behavior change, they are at best a "nice to have." The example from Medisafe in Figure 3.3 is not a compelling reason for a user to pay for an upgraded experience. A more meaningful one would be for users to choose how to communicate medication behaviors to care providers.

The choice that new Happify users are given to have their account in Private Mode or Community Mode *is* meaningful (see Figure 3.4). Some people feel behavior change is personal and prefer not to share their activities beyond their inner circle, while others thrive on social support. Letting users express their preference at the very beginning of their experience with Happify signals that their autonomy will be supported as they use the product.

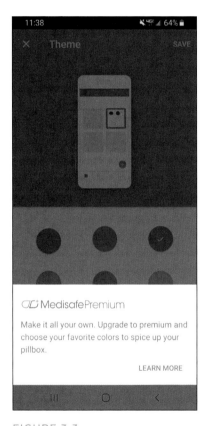

FIGURE 3.3
Medisafe's value proposition for users to upgrade is to be able to personalize the appearance of the app.

FIGURE 3.4
Happify has new users select whether they'd like their activity to be viewable by other users for support and encouragement, or if they want it to remain private.

What Really Matters

Meaningful choices often tap into people's core values and priorities. As product designers, you can help people think about what really matters to them and draw connections between their values and behavior change. Because not everyone has reflected deeply on what matters to them, you also may need to draw that out with thoughtful questions, prompts, or activities.

A technique that product managers sometimes use to understand what features are really crucial is known as *question laddering*. The idea is that rather than just accepting a certain feature is needed, the product manager asks, "Why is it needed?" They keep asking "why" until there is no further way to break down the answer. This allows them to figure out the true underlying purpose of the requested feature, so they can think creatively about the right way to achieve it, rather than thoughtlessly implementing only what was requested.

You can use the same question laddering technique to help your users think about why their goals are important to them. Users' first answers to a question about their goals may be very tactical. It's not useful for you to hear that they're using a weight loss app in order to lose weight. What you really care about is *why* they want to lose weight. Is it to boost self-esteem? Gain energy for keeping up with rambunctious toddlers? Stave off a health crisis?

The conversation about what really matters to your users might look something like what Noom does to help new users establish their goals, in Figure 3.5.

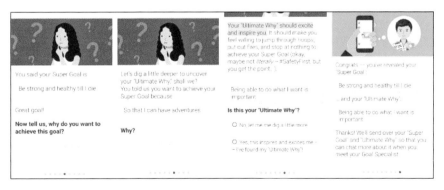

FIGURE 3.5
After Noom users describe their goal, the app asks them to dig deeper until they've gotten to the root reasons why that goal matters to them.

Whatever... I Do What I Want

People experience things they *want* to do differently than things they *have* to do, even if the activities are pretty similar. Remember *Farmville*, the Facebook game that swept the nation around 2010? At its peak, it had almost 84 million players who would log in every day to... plow their fields. Performing basic agricultural maintenance tasks like harvesting crops earned players experience points and game currency that enabled them to expand their farms (see Figure 3.6). Which, of course meant more farm chores to perform.

FIGURE 3.6
Farmville players enjoyed logging into the game every day to clear their fields of debris and fill their coffers with Farm Bucks.

Contrast that with a worker who goes to an office every morning and fires up the computer to review data in spreadsheets. Performing basic accounting maintenance tasks like updating revenue projections earns the worker their paycheck. While the worker may show up every day, they're not particularly jazzed about it, and spend most of the workday checking the clock in case it's time to go home.

Farmville and work have a lot in common. They both require daily effort, the completion of tedious and sometimes dull tasks, and they're impossible to "win"; there's always something more to do. Yet people were incredibly engaged with *Farmville* and experienced fun playing it, while there's broad evidence that many American workers are disengaged from their jobs. Could it be that when people elect to do something for their own reasons, it's more fun?

The Designer's Dilemma

There is a problem inherent in having our users' autonomy be a priority. Designers are rarely truly agnostic as to what their users decide to do. They have a vested interest in them making a particular set of choices over others. The success of their product counts on users doing certain things, whether it's spending their money on the product or achieving a behavioral outcome that leads a third party to spend theirs.

It's a leap of faith to design with the users' autonomy in mind because it means allowing for the possibility that users won't do what designers want—what designers need—them to do. It feels scary. That's okay.

Designing for user autonomy is playing a long game. If you get users to make the "right" choices with your product through force or trickery, you can achieve short-term outcomes. Most metrics won't tell you in real time if you've lost your users' trust. That lesson emerges down the road, when the problem may no longer be fixable. You'll discover you have pissed-off former users who are only too happy to tell their friends to stay away. Maybe even worse, you'll have people who tried behavior change but didn't stick with it and are now less likely to try again.

Consider automatic subscription renewals, which rely on people's forgetfulness and inattention to detail to make an ongoing profit. Sure, you can bury the renewal information in the fine print so that you get an ongoing payment for months, maybe even years after someone's stopped truly using your product. But what happens when the person finally takes a close look at their credit card statement and realizes how much they've paid you against how much value they've received?

It's not a good look.

That's not to say these sorts of auto-renewals are universally bad. They can be incredibly convenient for products like insurance or internet plans. But the key is that users should be aware that their payment will recur and have explicitly agreed to it. If they change their minds, it should be relatively simple for them to update the renewal.

Designing for user autonomy is scary. Driving users away forever because you didn't respect their autonomy is even scarier.

A Transparent Path to Change

Part of freely making a meaningful decision is understanding what that decision entails. People will agree to awful things if they believe it will help them achieve something they value. Think about what it takes to lose weight. People willingly agree to do things like:

- Stand on a scale and feel bad about the information displayed
- Ignore the growling in their belly that is begging for a delicious cupcake
- Move their bodies so much that they sweat through their clothes and need a shower and feel sore the next day
- Opt out of happy hour to go to a smelly gym instead

The mechanics of weight loss, when looked at a certain way, are really unpleasant. But people do them, and may even eventually take pleasure in them, because they believe that these actions will lead to an outcome they value.

How Not to Hook Users

Imagine that you want to lose weight and you google "best weight loss apps." The top result is for a program you've never heard of: Dr. Nick's Magic Diet. Dr. Nick looks a little sketchy, but he promises results in one simple step. You're in.

You fill out all of the information to enroll in Dr. Nick's Magic Diet: your birthdate, your address, your height and weight, your fitness goals, your health background, and finally, your credit card number. At each step along the way, the program encourages you: "You're almost there! Just a few more steps, and you'll receive the one simple step you need to lose weight!"

Finally, you reach the end of the sign-up and there's a big shiny button on the screen that says "Dr. Nick's One Simple Step." You are so excited. Here it is, the one simple secret to achieving the body you've dreamed of! You click, and the screen slowly fades to reveal the step:

> You've taken the one simple step to weight loss by joining Dr. Nick's Magic Diet. Now we'll get started on changing your diet, getting you into an exercise program, and looking at other habits you can change.

Wait a minute. You were promised one simple step, and now it turns out that there are really something like sixteen not-so-simple steps hiding behind the curtain. You're angry. This program is a joke. How can you get your money back?

This example is fictional, but the general principle is surprisingly common in digital products. Designers aren't always clear up front with users what they're committing to by enrolling in a program or using a product. That means their users don't have the opportunity to weigh the pros and cons of participating. Users can't make an informed decision about whether they're willing to put up with the drawbacks to gain the advantages of the product. And that means they're less likely to stick with it.

Tell the Whole Truth

Behavior change is hard. Most people realize that. If a goal is important to them, they are often willing to put up with some hardship. As a designer, it is your job to be truthful about what that hardship might look like within the experience you've built. That gives users who aren't willing to make those particular sacrifices the chance to walk away, while others who find the trade-offs reasonable will move forward.

As the designer, you may not always know what aspects of your behavior change design are positive or negative for a given user. Users will have different preferences and priorities. Some users might love the idea of going cold turkey on smoking cessation, while others prefer to use nicotine replacement therapy. Some may hate that a program has a social component, while others thrive on sharing their progress. When you describe what your behavior change program entails, keep it factual. Your users can decide for themselves whether the experience you describe is one they want to try.

Oftentimes, designers don't so much lie about what their programs include as they fail to be fully explicit with the truth. Take the example of Sweatcoin. Sweatcoin is a physical activity program that counts outdoor steps and converts them into currency. That's pretty much the language they use on their website to describe their program (see Figure 3.7).

After you start using Sweatcoin, you see how much currency you've earned that day right in the center of the home screen, and your balance at the bottom (see Figure 3.8).

FIGURE 3.7
Sweatcoin pays you in currency for the steps you take outside. Note they don't tell you which currency they pay with.

FIGURE 3.8
The center of the screen shows how much currency you've earned, both today and to date.

What wasn't apparent to me until the second or third day that I used Sweatcoin was that the currency they paid me with was not U.S. dollars. It was Sweatcoins. Sweatcoins can be redeemed for items in the app marketplace and even exchanged for real money once you pass an earnings threshold, but they are completely useless at the local mall.

To be clear, Sweatcoin never *says* it pays in dollars. If you scroll down on the product home page just below the main messages, they clearly tell you all about the Sweatcoin currency. But like many users, I didn't initially scroll down below the big call-to-action at the top of the screen. And the Sweatcoin symbol could be mistaken for a stylized dollar sign if you were just glancing and not looking closely, as I did. An important piece of information that should have been made very obvious to me was communicated subtly, and I missed it. And I work on this type of program for a living!

Be up front with your users (or potential future users) about what your product is and what using it will be like. You may find fewer people who are on board, but those who do get on board are more likely to stick with you over time.

Play by the Rules

Once users are enrolled in your program, it's important to provide an experience that aligns with what they expect. If you've promised them smooth sailing and then immediately hit them with challenges, they'll feel betrayed. Not only will they be skeptical about working on the challenges, but they'll also have negative feelings toward your product that wouldn't be there if they'd just known the requirements going in.

Often, designers fudge the truth to make their product seem more palatable to users. They might have rationales like the following:

- If people do a trial on the program, they'll like it so much that they'll be willing to upgrade to a paid version. It's okay not to tell them it's not free until they're already hooked.

- I know this is the best thing for users. I also know that if I tell them what they need to do, they won't do it. If I can get them started on the right path, I'll convince them.

- This is a really short survey. I'll say it's five questions, even though it's really ten; five sounds better, and no one will mind because it's still super short. (See Figure 3.9 for a program that did exactly this.)

The product designers who decide to present their users with more palatable experiences than they actually deliver aren't evil. They're just trying to get people to use their product and hoping it works out. It's not effective. Don't do it.

Give People an Out

If you heat water in a pressurized container with no steam valve, it will eventually explode. Likewise, if you repeatedly push your users into actions they don't want to take without an opportunity to decline, they'll eventually exit your product (and maybe badmouth it to all their friends). Unlike Don Corleone, behavior change designers shouldn't make offers that people can't refuse. When users can't opt out, their autonomy is not being supported. And without a way to opt out of a specific choice they don't want to make, people will take the ultimate opt out: they'll stop using your product altogether.

FIGURE 3.9

Stash says they'll ask five questions to suggest an investment portfolio for you. It's actually ten.

For the most part, behavior change designers need to have ongoing relationships with their users in order to accomplish the goals of their products. Upsetting users by cornering them on an action is counterproductive if the users don't stick with the program after that. If an action is not truly necessary for the behavior change program to work, let people postpone it, choose from a handful of options, or opt out entirely as they wish.

Letting people opt out may feel scary, but there's clinical evidence that it can plant the seeds for later action. There's a technique called "rolling with resistance" from motivational interviewing, which involves agreeing with people's negative statements. "Maybe you're right, it doesn't make any sense to quit smoking." The human brain is wired to protest against these types of statements, so that sometimes gets people thinking differently. "Wait a minute, I can quit if I want to!"

Maybe Later

Sometimes people are not ready to take a particular behavior change step. A general good practice is to respect people's level of readiness. Behavior change usually involves a number of steps. If your user is ready for Step 3, then offer Step 3 and maybe present some information about Step 4 to get them thinking about the next part of the process. Talking about Step 8 is more likely to be overwhelming than helpful.

You won't always know what your user is ready to try, so consider giving them an opportunity to postpone an action instead of just opting out. If they opt out, they may miss out on trying that action when they are ready later. A "maybe later" option also helps your design team understand that users are not really rejecting that option, but that it may be presented at the wrong point in time for them.

An example of "maybe later" in 23andMe comes with seeing some genetic reports that could potentially be upsetting to users (see Figure 3.10). These reports concern genetic predispositions to serious health conditions. Instead of just displaying users' results on these dimensions the

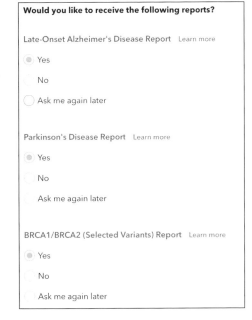

FIGURE 3.10

23andMe users can choose to postpone seeing the results of certain genetic analyses if they don't feel ready or interested to look now.

same way they reveal information about users' hair texture and ear wax type, 23andMe asks them to opt into seeing them now, later, or never. If users do decide to see the report, they also have to complete a quick learning module about how to interpret the results before they can see them. This design is intended to present potentially difficult information to users only when they are ready and prepared with the information they need to understand it.

Door Number 3, Please

Another way to give people an out is to offer them options. Any time someone picks one option out of several, they are implicitly saying no to the options they don't choose. Multiple choice structures let people say no without saying no.

Consider this example from Noom. Noom has a live coach as part of its programming. During onboarding, the app asks users to indicate their preferred way for the coach to encourage them (see Figure 3.11). If a user selects "validation and acknowledgement" only, it's implicitly telling the coach not to send motivational quotes.

When There Really Is No Choice

If something is really critical to a program's success, provide the rationale for why and make it easy for users to accomplish. Some examples include:

2. How can your Group Coach best help you navigate the ups and downs of your weight loss journey?

○ Motivational quotes.

○ Supportive feedback.

○ Practical tips.

○ Group challenges.

○ Validation and acknowledgement.

○ All of the above.

FIGURE 3.11

Noom users can implicitly opt out of certain types of encouragement from coaches by choosing a different option instead.

- A program that teaches budgeting behaviors requires users to input financial details before an initial strategy can be created.

- A health assessment of genetic risk needs users to submit a saliva sample before any information can be provided.

- An exercise program for people recovering from serious injury needs to be approved by the user's physician to make sure it's appropriate before it can start.

Some behavior change programs require users to own specific hardware beyond a computer, tablet, or phone. Garmin Connect only works if you use at least one Garmin device; Shapa needs you to step on a Shapa scale; and the ICAROS Home "personal active virtual reality system" requires a headset, a large workout frame, and a willingness to look very silly (see Figure 3.12).

FIGURE 3.12
The Verge calls working out on the ICAROS Home "a cross between VR sky-diving and planking." If you want to try the app, prepare to pony up thousands of dollars for hardware.

Align Choices with What Really Matters

Another tactic to help people make a decision is to highlight how options *do* or *don't* align with their values and goals. In general, people will find an option more attractive if it reflects the things they care about.

You may remember learning about cognitive dissonance in Psychology 101. When people behave in ways or have attitudes that are out of step with their values and they know it, it feels lousy. It's like an itch they can't scratch. *I'm an honest person ... why did I tell that lie?* People feel more comfortable with themselves when their attitudes, beliefs, and behaviors are in harmony with their goals and values.

Most people don't like experiencing emotional discomfort. When there's misalignment between attitudes, beliefs, behaviors, and goals and values, they will change one of the elements to restore harmony.

TIP VALUES ARE THE ANCHOR

It's a lot easier to change an attitude, belief, or behavior than it is to change a value or a deeply meaningful personal goal. That means you can use those values and goals as launch points to drive behavior change.

Getting people to see how their behaviors don't support their values or goals takes a skilled touch. If you bluntly tell someone their behaviors aren't value-consistent, they're likely to take offense. If you're lucky, they'll just stop listening to you instead of getting openly angry. Unless you have a close and personal relationship with someone, you can't just march up to them and accuse them of hypocrisy.

Here are a few better ways to approach it.

Motivational Interviewing

Motivational interviewing gives us a conversational tool we can use to gently point out inconsistencies between people's beliefs, attitudes, behaviors, and the things they value. It's a clinical technique, but it adapts really nicely to all sorts of situations. Best of all, you can borrow its language in digital products, as well as face-to-face interactions.[1]

I like to refer to motivational interviewing as "conversational judo." In martial arts, you don't exert your own force so much as work with the other person's force. Similarly, motivational interviewing is about giving voice to a person's values and any inconsistencies they're experiencing, and letting the other person react. See the example in Figure 3.13, from Change Talk, an app that teaches motivational interviewing techniques to healthcare professionals.

---

1 Research suggests that motivational interviewing is most effective in face-to-face coaching, but that using these tactics in digital interventions can prompt change talk from participants. In a digital format, you're borrowing tools from motivational interviewing rather than practicing it wholesale.

FIGURE 3.13

Change Talk teaches healthcare professionals to prompt patients' change talk by reflecting back their values and letting them sit with the inconsistencies in their behaviors. The example in this screen is a mother who is reluctant to limit her overweight son's soda consumption.

When you use motivational interviewing, you don't try to argue with someone or convince them of your viewpoint. You simply raise the inconsistency and ask them to react. Here's what an example conversation using motivational interviewing might look like in a smoking cessation program. (Yes, some of the newer conversational agent technology can handle these sorts of conversations!)

Program: Thanks for completing your personal profile, Kevin. We can use this information to help develop a quit plan that's right for you. One thing that stands out is that you have three kids and say spending time with them is one of your top priorities. Is it okay to ask you to think about how you balance smoking with having the energy and focus for your kids?

Now Kevin might read this and tell the program to butt out. That happens. But what also might happen is that this idea starts to creep into Kevin's thoughts. It's that itch he can't scratch. The question has planted a doubt in Kevin's mind, and it's gradually making him feel less and less at ease.

What happens next? Well, Kevin is going to try to reduce his discomfort. One option will be to downplay the importance of being a dad, which his co-parent and his kids both hope he doesn't do. Another is to seriously work at quitting smoking. As hard as that is, it's a hell of a lot easier than abandoning his values.

What can you do to infuse the conversational judo of motivational interviewing into your digital programs? Here are a few tactics:

- **Use rhetorical questions.** Since motivational interviewing isn't about arguing, it's not important that you have a response ready for people. Stating the inconsistency and asking a question like, "How do you explain these two things both being true?" will do.

- **Avoid compulsory language.** When you fight people, they often double down. It's a natural reaction to threat. So don't tell them what they "should" or "need" to do.

- **Don't tell people that they are wrong or cast explicit judgment on them.** Don't tell Kevin he's a lousy father. You have no basis for that judgment, and it's just going to shut Kevin down.

- **Ask your users to tell you about their values and personal goals when you're learning about who they are.** Even getting a high-level glimpse will give you something to work with. Once you've gotten some feedback, you can reflect back to people what you know about their values, especially in juxtaposition to the behaviors you're working together to change.

- **Call on universal human values.** Research shows that across cultures, ages, and other groupings, there is a core set of things that tend to be highly valued by people. They include community, family, security, and the natural world. If you don't know a lot about your users personally, reflect values that they are likely to have for them to consider alongside their behaviors. Even if Kevin hadn't told you how he felt about being a father, it's a pretty smart guess that he wants to be a good one.

One of the ultimate goals of motivational interviewing is that change talk comes from the person making the change, not from a coach or expert. So restrain yourself from laying truth on your users; instead, think about creating conversational breadcrumbs that lead them there.

Package Values with Choices

Complex choices can be simplified by describing them in terms of the values they support. Similar to how ecommerce sites tell you "Other people who bought this item also bought these three items," value-aligned packaging says "People who care about this particular cause support it by making this choice."

An example of a product that does this is Stash, an app that aims to change people's behaviors around money management by getting them comfortable with investing. New users can complete a quiz that assesses what they value most in their investments (see Figure 3.14). The options mix personal values like "social responsibility" and "environment" among more traditional investment features like "dividends." Then Stash recommends investment funds that align with the selected values (see Figure 3.15). Or, users can

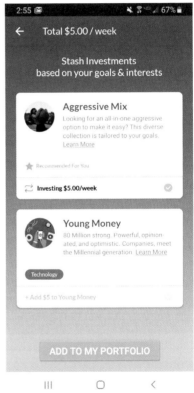

FIGURE 3.14
Stash lets users indicate what's important to them before recommending specific investment products.

FIGURE 3.15
After Stash learns what users value, it suggests a few fund options that align with those values.

browse investments by the causes they support (see Figure 3.16). By reframing what might be a complicated choice for someone without a background in investing as a values-based decision, Stash simplifies the decision process and eliminates one barrier to investing.

FIGURE 3.16
Users can also browse Stash's fund offerings by cause and click to choose investments that support their top values.

Use Choices to Reflect Values

Sometimes, people like to engage in what's known as *virtue signaling*. This is when they do something publicly that demonstrates something positive about themselves. Someone who posts the details of their run on social media right after they get home could be said to be virtue signaling about their fitness.

The phrase "virtue signaling" implies an element of performance or a lack of sincerity, but there are reasons why someone might want to broadcast their own positive qualities. People like to express their values to others. They believe that the things they value are good. When they do something that shows their values, they expect that others will view them positively for it. So sharing evidence that behaviors match values can feel rewarding for people. Behavior change programs that let users share their value-relevant accomplishments will reinforce their decisions.

One example comes from NOAA Planet Stewards Education Project (PSEP), which trains educators to teach about sustainability and encourage green behaviors from students. One of their training programs was a personalized learning game where users could specialize in a specific area of conservation such as Climate, Marine, Freshwater, or Oceans and Coasts. As learners completed activities, they could earn digital badges to add to their online profiles that signaled their accomplishments (see Figure 3.17). These badges served a dual purpose of providing credentials and showing what the person valued.

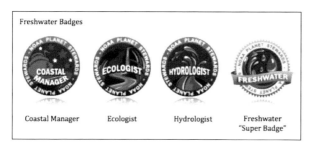

FIGURE 3.17
NOAA's PSEP badges reflect not just learners' accomplishments, but also what they value.

Don't Go Overboard

It takes a deft hand to successfully get users to consider their values and behaviors in tandem. If you come on too strong, people may feel angry or manipulated.

Ever go to a website and get interrupted by a pop-up asking for your email address? More and more, those pop-ups include a value proposition intended to make signing up for the site's email list more appealing.

This is a case of designers badly misusing a tactic rooted in behavior change principles. What these pop-ups are trying to do is to get users to realize that signing up for the email list aligns with something they value, and therefore they should do it. What they're missing is that it's very rare that a meaningful value is truly supported by inclusion in a marketing database. Consider the examples in Figure 3.18. How many of these value propositions really make you think sharing your email will help you achieve a meaningful personal goal?

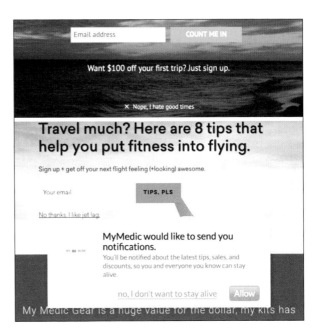

Being too heavy-handed with value alignment will backfire. Your
users are smart people. Save the heavy talk about life purpose and
personal goals for when it actually makes sense.

The Upshot: Let There Be Choice

This chapter has focused largely on how to offer users meaningful
choices. Specifically, users should be able to express their personal
values and goals along the behavior change journey. When there are
opportunities, the behavior change designer should look to either
align behaviors with those values and goals, or highlight how the
behaviors, values, and goals are connected to each other. Maintaining
a strong tie between what users are doing and what is most meaningful to them will help sustain behavior change over time.

Of course, it's not quite as easy as putting a choice in front of a user
and watching magic happen. People often need help and support to
make the right choices to reach their goals. In the next chapter, you'll
learn how to design experiences that will help people arrive at good
choices that support what matters to them.

If it weren't for Vic Strecher, you probably wouldn't be reading this book right now. My first behavior change design job was at a company Vic founded, HealthMedia. I found my life's calling working on HealthMedia's digital health-coaching products. I also learned about how life purpose can be a path to change, which is Vic's professional and personal focus.

Vic realized the value of purpose in the aftermath of a tragedy, when he lost his 19-year-old daughter Julia to a rare heart disease. In his search for a way forward through grief, Vic realized that purpose has transformative powers. Yet, it was completely missing from conversations around health behavior change. In 2015, Vic founded Kumanu to help people connect authentically with their purpose and use it to power positive change in their lives.

How do you make a complex idea like purpose accessible?

Ask easier questions up front. Start with questions like, "What are you like when you're at your best? What are the things that matter most? Who relies on you?" Look at what's on the wallpaper of your smartphone. Often, we display things that matter a great deal to us. Or try the "headstone test": "What would you want imprinted on your headstone? What do you want people to remember or say about you?"

This is a process over time. Ask existential, thoughtful questions over a period of days. Then after eight or ten days or so, start asking about people's purpose and allow them to refine it. You continually build a more authentic purpose this way.

Once people have purpose, how does it drive behavior?

How do you live if being purposeful means applying your best self to what matters most? How do you become your best self?

That becomes the basis of what we start working on with a person. As Nietzsche said, if you have a why, you can accomplish any how.

I ask people to consider what they're like when they're at their best. Let's say my best self is being calm, and part of my purpose is I'm going to be calm with my employees. In that case, being calm becomes a target for my activities. Kumanu's Purposeful app then assembles suggestions for small habits that can help me with this goal, maybe through meditation and mindfulness exercises, physical activity, or eating right through the day. There are quite a few things Purposeful can do to help a person become their best self day in and day out. We use machine learning to start identifying things that seem to be working for people and continue to get smarter. Netflix uses machine learning to figure out what movies you like. Purposeful uses machine learning to figure out how to help you become a better human.

How does purpose change the brain?

Thinking purposeful thoughts stimulates the ventral medial prefrontal cortex, the part of the brain that relates to identity. The amygdala is the reptilian part of our brain, and relates to fear and aggression. When we try to help people make changes in their lives without the self-affirmation of purposeful thinking, they often react defensively. The amygdala often starts flipping out, and you're stuck with fear and aggression. The amygdala, however, is regulated by the ventral medial prefrontal cortex, so purposeful thinking is a great exercise to begin reducing defensiveness to change, becoming more open to change, and relating the change to whom you are in a very deep way.

A recent study that we conducted also showed that people who have a strong purpose have less activity in the dorsal anterior cingulate cortex (DACC), a part of the brain that relates to conflict. The DACC starts getting more active when you go, "Oh man, should I play with the kids, or have that Old Fashioned?" If you have a stronger purpose, you're less conflicted. You know what to do.

How long do people need coaching to live purposefully?

Think about self-help and what self-help is really about. If they're really engaged in a self-help program, people should be able to leave your app. They should be able to leave your program as an improved person, as a self-determined problem-solver who's able to self-manage and doesn't need you anymore. That would be great self-help!

I really think that the ideal program is where a person uses it and improves and then they go, "Wow. I guess that app was nice, but I really did this on my own." So the attributions for change are actually internal as opposed to external. I feel that some of the best self-help programs, you don't even attribute your success to. Some people may use our app very briefly and really get a ton of effect from it. The proof is in the pudding. Did people make significant changes in their lives? Whether they attribute it to you or not, who cares?

Victor Strecher, Ph.D., M.P.H., *is a Professor at the University of Michigan's School of Public Health and the founder and CEO of Kumanu. Prior to his current roles, Vic founded the U-M Center for Health Communications Research, as well as HealthMedia, which was acquired by Johnson & Johnson in 2008. Vic has written several books, including* Life On Purpose: How Living for What Matters Most Changes Everything, *and* On Purpose: Lessons in Life and Health from the Frog, the Dung Beetle, and Julia.

CHAPTER 4

Weapon of Choice

Make Decisions Easier

In the last chapter, you learned that it's important to give your users meaningful choices to make on their behavior change journey. In this chapter, you'll learn how to structure those choices so that it's easier for people to select good options that ultimately support their goals.

The fact is, people aren't great at making decisions in real life. They have limited brainpower and short attention spans that make it hard to sift through information and make sense of it. The shortcuts the brain has developed to cope with this issue aren't perfect and lead people to a predictable pattern of mistakes in judgment. People may also find themselves in a struggle between heart and mind, when what they most want to do isn't what they think they should do to reach their goals. Decisions aren't just logical. They affect feelings, too.

To make matters worse, some designers deliberately take advantage of the flaws in human decision-making to push people toward certain courses of action. Whether it's offering a too-good-to-be-true introductory rate with the expensive monthly subscription buried in the fine print, or deliberately stacking one choice with positive language that biases decision-makers, designers have all kinds of tricks to nudge their users down one decision path. For effective behavior change that sticks even when people stop using your product, trickery is not the answer!

Instead, you can design an experience that both gives people accurate information about their choices and helps them pick one that aligns with their goals. Once you understand why good choices are hard for people, you can use your designs to make decisions *easier*.

In the process, you'll be doing your users a lot of good. Guiding people to better choices can make them happier. Research shows that happiness is at least partly determined by what people pay attention to. If they focus on life's positives, they're more likely to feel happy. Designers can build a focus on the good into the choice architectures they create to infuse behavior change with a little more happiness, while leading people toward their behavior change goals.

Why Choices Are Hard

You know that giving people choices helps them commit to long-lasting behavior change. But, simply giving them a buffet of options tends to backfire. People need some guidance and structure in order to arrive at a good decision. Designing an experience that offers people the opportunity to choose *and* makes it easy for them to make a good choice requires understanding the reasons why people are so lousy at decisions in the first place.

The Paralysis of Analysis

Variety is the spice of life, right? Not always. Too many options can make it hard for people to make a decision. More decisions mean more data to mentally crunch and more risk of picking the wrong option. So people freeze up. That state of indecision in the face of lots of options is sometimes called *analysis paralysis* or *the paradox of choice*.

In hundreds of studies, researchers have seen that when there were more options to choose from, people had a harder time making any decision at all. A digital version of option overload that may be familiar is Netflix (see Figure 4.1). Sometimes it takes longer to find something to watch than it does to watch it.

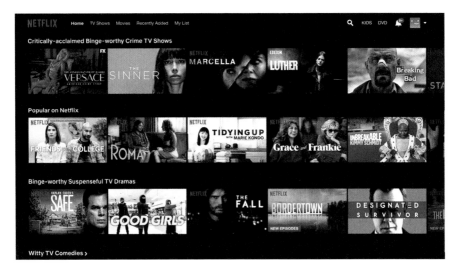

The streaming options on Netflix seem endless, which actually makes it harder to choose something to watch.

So while it may seem counterintuitive, it is helpful for users to have *fewer* options rather than more when they are being asked to make a decision about their behavior change process. Too many choices can overwhelm people; they may feel like they don't have control, or that they *have* to pick the perfect option from thousands. These feelings are the opposite of autonomy support. The job of the designer is to help people break down complex decisions into meaningful choices.

Disorganized Data

People are pretty good at making decisions when they have equivalent data on a few options. For example, it's not a problem to pick between these three options:

- A one-month subscription for $10

- A three-month subscription for $28

- A six-month subscription for $50

There are only two variables—duration and cost of the subscription. People can consider how long they might want to use the product and how much they can afford to pay and use those two factors to make a decision. Easy.

But as you well know, most real-world decisions aren't nearly so simple. Consider the decision to purchase a smart scale. One person might be interested in buying a scale that integrates with her favorite fitness app. Another person really wants one that can assess body fat and water weight in addition to just weight. Each of these people might be willing to concede some of their desired features for the right price or if there were an attractive alternative feature they hadn't thought about.

Both people take to Amazon to see if they can find the smart scale of their dreams, and they see the following chart comparing top recommended options (see Figure 4.2). The person who wants to use her app can see that two of the scales integrate with *an* app, but not which app or if there are multiple app options. So she ends up clicking into all four descriptions and hunting for the information. The second person also needs to do some investigating to verify which scales do and don't work for him. Does the RENPHO include water weight in that body composition feature? And that first Etekcity scale looks pretty high tech—could it do body composition, too?

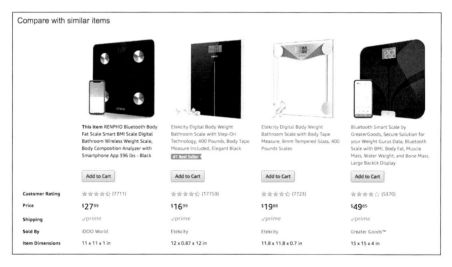

FIGURE 4.2

Amazon devotes some of its limited real estate in the comparison chart to the physical dimensions of the scale, but none to the specific technology integrations each scale offers.

Hunting this information is time consuming, makes the decision more difficult, and may ultimately dissuade people from making any decision at all.

Sometimes the data disorganization is a deliberate ploy by designers to influence people to pick the most profitable option. The design might strategically highlight or withhold information so that one option appears superior to others. Or information might deliberately be presented in a confusing way so that people default to a familiar choice.

While using design to nudge people toward good choices can be a helpful strategy, it's important not to do it from self-serving motivations. Try not to deliberately make comparisons difficult for people. Some designs either withhold information about some options to make another one look better, or have information organized so that it's difficult to compare options accurately. This is dark side type of stuff.

The example from Lifesum in Figure 4.3 recommending a ketogenic diet leaves out key information that would help users decide whether they should follow that advice. It highlights benefits and teases the appealing outcomes of a lower body fat percentage and restored metabolism without having assessed whether the diet is appropriate for a given user. Worse, it doesn't give users any prompts to make that assessment themselves.

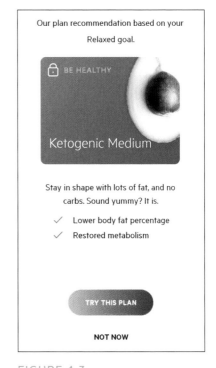

FIGURE 4.3

By providing only benefits with no drawbacks or selection criteria, Lifesum probably increases the number of users who say yes to trying a ketogenic diet, but to what end?

Confusing Complexity

Sometimes the problem is not nefarious designers, but truly complicated decisions. Consider someone who's looking at medical options to treat a health condition. Unless the patient is also a doctor, there will be unfamiliar terminology, misunderstood trade-offs, and probably uncertainty about which option is best. A situation many people experience is selecting health insurance coverage or retirement plans, but who really understands in detail all of the factors that go into each option? Many consumer and behavior change experiences ask people to make a decision with far-reaching consequences, without the deep understanding that would help them truly optimize.

These problems of complexity are compounded by literacy issues. Health literacy is people's ability to obtain, process, and understand health-related information. The National Institutes of Health estimates that about 90 million American adults have poor health literacy. Then there's numeracy, the ability to apply simple mathematical concepts and understand quantitative information. The statistics for numeracy are even worse in the U.S. than for health literacy. If people don't have the basic skills to comprehend health information or numbers, then they are at a huge disadvantage in making complex decisions that affect their well-being.

Complexity affects behavior change domains beyond health. From deciding how to prepare for retirement to evaluating career training options, some of the most meaningful behavior change journeys people take require them to choose from complex options. Far too often, people are not equipped with the skills or support to do that well.

Opportunity Costs

People are lousy at considering opportunity costs associated with the choices they make. Every course of action requires people to spend resources: time, energy, and possibly money. Every resource spent on one action is no longer available to spend somewhere else. Even the few seconds it takes people to become aware of a reminder pulls their attention away from other tasks.

Almost always, deciding to take one course of action means *not* taking alternatives. People may not fully think through the implications of the choices they make and find themselves surprised that saving money toward a down payment means not having the cash for Friday cocktails.

Most of us know *willpower* as something we wish we had more of. Will-power gets portrayed as a transfixed quality that can make people "badass doers" if they have it, or "lame wannabes" if they don't. But willpower is not a personality characteristic like a sense of humor. Science shows it's more like a muscle.

Successfully building muscle requires working out to fatigue the muscle and then taking a rest so the muscle can repair itself. Over time, the muscle becomes stronger from this cycle of work and rest. Similarly, willpower can be depleted from being overused. And, it can be recharged through periods of rest. If done strategically, a person's store of willpower can become greater over time, just like a muscle gets stronger.

What sorts of things are willpower workouts? Making choices! Having to decide between multiple options is an emotionally and mentally stressful situation. Give someone too many choices, and eventually they won't have the focus to make good ones. People's minds can suffer from *decision fatigue.*

Think about what it's like to be on a diet and have to turn down the tempting unhealthy options at every meal. It's not so bad at breakfast and lunch, but by late afternoon, those decisions get really hard. Fatigue can occur even within a single flow. That's why it can be a good idea to put hard survey questions first, when people will have more willpower to process them, or to simplify checkout after people have done the hard work of selecting a product.

When you make decisions easier for your users, you're helping them preserve their willpower to use in the situations where they need it most.

Your Brain Is Biased

Unfortunately, the human brain is not a perfect logic machine. It comes preloaded with cognitive bias software that shortcuts the decision-making process, but makes a predictable series of errors in doing so. By some counts, there are over 180 cognitive biases that color the way that people interpret information and make decisions.[1]

Not every cognitive bias will come into play for your users and product, so there's no need to try to learn all 180 of them. Figure 4.4 summarizes some of the more commonly encountered biases so that you can be aware of them.

---

1 Buster Benson/Better Humans and John Manoogian III created a visual diagram called the *Cognitive Bias Codex* that can serve as a helpful reference for designers.

Cognitive Bias	Description	Example
Anchoring	The first option a person sees *anchors* their choice.	Sarah signs up for fewer volunteer sessions when the default suggestion is 1 than when it is 4.
Availability bias	People believe things they can remember more easily are more common than they really are.	Joe remembers a celebrity lost weight on a juice diet, so thinks juice diets are healthy and normal.
Confirmation bias	People seek information that confirms what they already believe.	Trey is sure his BMI is high because he's so muscular, so he discounts the body fat readings on his smart scale.
Loss aversion	People are more sensitive to losing what they already have than gaining something they don't have yet.	Maria maintains her gym membership even though she never goes because she knows she'll never find another one this affordable.
Present bias	People are more attuned to what's happening right now than what happened in the past or will happen in the future.	Lamont decides to spend his bonus now instead of investing it in an interest-earning bond that will mature in 3 years.
Status quo bias	People will stick with what they know over an unfamiliar option.	Teresa is unhappy with her job, but hasn't looked for a new one because she's not sure it will be better.
Mere exposure effect	Once people see information somewhere, they're more likely to believe it's true when they see it again.	After seeing anti-vaccination posts on social media, Kyle could swear there was one study that linked the MMR vaccine to cancer (nope).
Sunk cost fallacy	People have a hard time walking away from an unsuccessful strategy if they've spent time, energy, or money on it.	Penelope hates her degree program, but since she only has a few semesters to go she makes the next tuition payment.
Outcome bias	People judge a process as good if the outcome was favorable.	David highly recommends the doctor who helped him rehab his now-healthy knee, forgetting how he cursed the doctor after every painful PT session.
Recency effect	People are more likely to remember things that happened recently and give them disproportionate importance.	Max didn't know anyone at the party and spent most of it wanting to leave, but met someone he really liked toward the end. Max remembers having a great time at the party.

FIGURE 4.4

These are some of the more common biases your users may experience as they work through behavior change.

No Choice Is a Choice

One of the cognitive biases in Figure 4.4 is the status quo bias, where people actively prefer the current state over an unknown alternative. Some people may stick with the same old choices, not because they want to, but because they're avoiding making another choice.

There are many reasons why people might make no choice at all. People avoid making decisions that are difficult or confusing, especially if they don't see a clear purpose in taking action. If the actual choosing process looks time-consuming or troublesome, they might give up on it. Unfortunately, *no choice* often *is* a choice to remain with the status quo, whether it's desired or not.

When "Good" Choices Are Bad

People are sensitive to information that tells them how well they're performing. One consequence of that is that people are vulnerable to what's called the *social desirability bias.* People have a sense of what the "right" or "good" course of action might be, and will sometimes do that instead of what they really want so that others think more highly of them.

People are most susceptible to the social desirability bias when a situation has a lot of *demand characteristics* that signal what they should do. For example, if people know their data is being monitored, they're more likely to do what they think others want instead of what they want themselves.

People opting into good behaviors for the wrong reasons may not seem like a problem until you remember that having ownership over a decision is crucial for withstanding challenges. When you're talking behavior change that needs to happen over time, it's important that people opt in for personally meaningful reasons. Otherwise, they're more likely to revert to their old ways the first time the new ones get too tough.

Making Choices Easier

Designers can make it easier for users to make choices. By creating *choice architectures* that work with, not against, the way people's brains handle decisions, designers can facilitate feelings of ownership, empowerment, and capability in their users. Here are some tactics for designing effective choice architectures.

Constrain the Choice

The behavior change designer is also a curator. For every target behavior, there are a universe of options that could help someone perform it. The designer should select a well-vetted subset of that universe of options to present to their users.

The trick to curating options is *to only offer good choices.* If a choice will take users astray of their goals, why tempt them into taking it? Use your subject matter expertise about the target behavior to identify a handful of good options for users to consider—the most practical, promising, or engaging ones you have.

Roobrik is a decision support tool for people considering assisted living options for themselves or loved ones. After users provide information about the person requiring care, Roobrik summarizes the person's needs and offers a limited number of options to pursue (see Figure 4.5). By curating the options for users—and providing a clear rationale for why those options rose to the top—Roobrik helps

FIGURE 4.5
Roobrik selects a small number of care options based on information the user has shared and then guides the user in researching each one.

to simplify the complex and emotionally fraught experience of choosing long-term care options.

Another way to constrain choices is to sequence them. Instead of having a user lay out their entire behavior change plan on Day 1, ask them to make an initial choice and take some action. As they progress, you can offer them additional choices to make. In the Fabulous app for "healthy rituals," the first decision is simply whether or not to drink more water for the first three days (see Figure 4.6). The designers use that first simple choice to build confidence and introduce the program structure so that later choices make more sense to users.

Why am I doing this?

I know, I know Amy! Setting a goal to drink water just feels too simple.

But we have a plan for you. We're starting very small, so we can build strong foundations. Soon, your challenges will rise to caring for your physical and mental health.

OK, GOT IT

WHY AM I DOING THIS?

I ACCEPT THE CHALLENGE

FIGURE 4.6

Fabulous users make one simple choice after onboarding—to up their water intake for three days. Once they accomplish that easy goal, they'll be asked to make meatier choices.

Create Symmetry

Compare apples to apples! Organize information so that people can compare options more easily on the most important dimensions. Distill the essential factors that people need to know to make a choice and focus only on those. Provide the same information for all options so that people can make an even comparison. Clear and consistent labels can help people navigate their options and compare like items, especially if information is presented across multiple pages.

On their site, Aflac helps people select a health insurance plan by comparing four options on important criteria, like how much it will cost to have that plan (Figure 4.7). Clearly, they have not included all features of these plans in the comparison grids. Instead, Aflac focused on some of the key factors that people consider in choosing a plan. Interested shoppers can go a level deeper to investigate additional details.

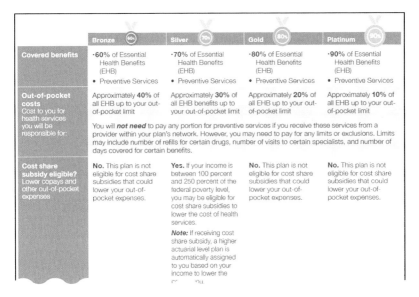

	Bronze (60%)	Silver (70%)	Gold (80%)	Platinum (90%)
Covered benefits	•**60%** of Essential Health Benefits (EHB) • Preventive Services	•**70%** of Essential Health Benefits (EHB) • Preventive Services	•**80%** of Essential Health Benefits (EHB) • Preventive Services	•**90%** of Essential Health Benefits (EHB) • Preventive Services
Out-of-pocket costs Cost to you for health services you will be responsible for:	Approximately **40%** of all EHB up to your out-of-pocket limit	Approximately **30%** of all EHB benefits up to your out-of-pocket limit	Approximately **20%** of all EHB up to your out-of-pocket limit	Approximately **10%** of all EHB up to your out-of-pocket limit
	You will **not need** to pay any portion for preventive services if you receive these services from a provider within your plan's network. However, you may need to pay for any limits or exclusions. Limits may include number of refills for certain drugs, number of visits to certain specialists, and number of days covered for certain benefits.			
Cost share subsidy eligible? Lower copays and other out-of-pocket expenses	**No.** This plan is not eligible for cost share subsidies that could lower your out-of-pocket expenses.	**Yes.** If your income is between 100 percent and 250 percent of the federal poverty level, you may be eligible for cost share subsidies to lower the cost of health services. ***Note:*** If receiving cost share subsidy, a higher actuarial level plan is automatically assigned to you based on your income to lower the c̶ ̶ ̶ ̶ ̶ ̶ ̶ ̶ou.	**No.** This plan is not eligible for cost share subsidies that could lower your out-of-pocket expenses.	**No.** This plan is not eligible for cost share subsidies that could lower your out-of-pocket expenses.

FIGURE 4.7

Aflac presents insurance options in a grid that permits easy comparisons on the characteristics shoppers are likely to care about.

Expose Just Enough Detail

Think about what level of detail to include in each option. If people are experts on a topic or care about details, then a lot of information will be helpful. If they're not equipped to parse through details, it will feel overwhelming instead of empowering. Imagine that you need a medical procedure. You may want to pick your surgeon and indicate where the incision will be, but few people care what brand of sutures is used.

A beautiful thing about digital experiences is that you can personalize the level of detail so that more advanced or knowledgeable users get more complex choices, while newbies get simplified ones. If you don't have the data to present users with the right level of complexity, ask them whether they're interested in a basic, intermediate, or advanced experience. You can also design an experience that works for the people with the least knowledge, while offering more advanced users some opportunities to dig deeper. For example, knowledgeable investors can look under the hood of the personal values–based investment funds offered by Stash that were in the last chapter, as seen in Figure 4.8, but novices can put money in them without perusing the detail.

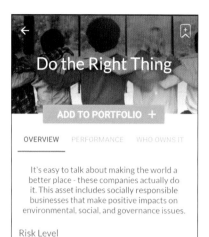

Do the Right Thing

ADD TO PORTFOLIO +

OVERVIEW   PERFORMANCE   WHO OWNS IT

It's easy to talk about making the world a better place - these companies actually do it. This asset includes socially responsible businesses that make positive impacts on environmental, social, and governance issues.

Risk Level

Conservative

Top Holdings	% of Total
MICROSOFT CORP	5.14%
ECOLAB INC	4.12%
APPLE INC	3.51%
3M	3.02%
ACCENTURE PLC-CL A	2.66%
ALPHABET INC-CL A	2.65%
BLACKROCK INC	2.08%
SALESFORCE.COM INC	1.87%
NORTHERN TRUST CORP	1.80%
AGILENT TECHNOLOGIES INC	1.76%

Investment Name

iShares MSCI USA ESG Select Fund

Ticker

SUSA

Last Price	Dividend Yield	Expense Ratio
$118.16	1.62%	0.25%

Visit Fund Website

You can also create bundles that combine many choices into one or two key decisions. This allows users to decide on themes or top qualities they want in a choice without having to repeat the process over and over. Weight Watchers does this for calorie counting. Their points system (see Figure 4.9) allows users to track just one simplified metric (points) instead of many.

Provide Structure

The way that you structure tasks and the decisions they involve for your users can help with navigating them successfully. Simple organizational tools like checklists or outlines can be extremely effective at moving people through multiple complex steps. Checklists of all of the steps involved in complicated tasks like surgeries or landing planes on aircraft carriers have been shown to increase reliability and decrease errors by a large margin. Giving your users a series of explicit steps to follow goes a long way in guiding them through behavior change.

A checklist or other structure can help with decision-making because it assigns the decision to a particular point in time, with the activities already completed and the ones yet to come clearly indicated. Checklists provide *context* and *set expectations* that help users make better choices.

Shared decision-making tools are one type of modified checklist being used in health care to guide doctor-patient discussions about treatment options. The example from Healthwise in Figure 4.10 helps patients decide about whether they want to receive CPR and life support if needed. After patients complete a step-by-step process, the tool generates a summary they can then use to talk to their doctor and make their desired arrangements.

In the previous chapter, you learned that personally meaningful choices are more likely to lead to long-term change. Research suggests that when people use these sorts of decision-making tools for their health decisions, they end up with an option that supports their values better. Making choices easy can help make them meaningful, too.

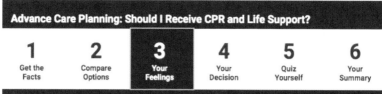

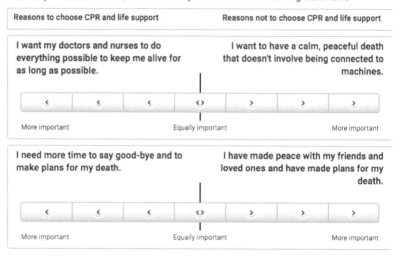

FIGURE 4.10

Patients who use Healthwise's decision tool walk through a checklist of steps to make a choice. The tool provides them with summary information they can share with their medical professionals.

Talk About Trade-Offs

People aren't good at understanding the trade-offs associated with their decisions. Your designs can make some of those trade-offs explicit so that it's easier for users to include them in the choice process. Consider this example from BecomeAnEx, a smoking cessation program developed by the Truth Foundation (see Figure 4.11). They've created short videos about different nicotine replacement therapies that talk about the pros and cons of each one. Users can decide for themselves which ones to try, with full awareness of the trade-offs they might experience.

Quitting Medications: Nicotine Nasal Spray

Nicotine spray gets nicotine into your system faster than other medications. It can be a little more difficult to use, but some tobacco users like to start off with it because it helps fight those cravings more quickly. The spray also requires a prescription.

FIGURE 4.11 BecomeAnEx provides information to help users make a decision about which quitting aids to use based on their advantages and drawbacks.

Another way to make trade-offs clear is to help users see "if, then" consequences of their options. Framing choices in terms of future consequences can cut through the vagaries of technical details to help people focus on what they care about most. The "debt snowball" is a technique for paying off multiple debts (hello, student loans!) efficiently. There are downloadable spreadsheets (see Figure 4.12) that help users enter the details of each of their debts, including interest rates and minimum payment amounts, and then experiment with allocating their monthly budget differently among the accounts to see which method yields the best results. Some people like to pay off the largest or highest interest debts first; others have more success if they can knock off a few smaller accounts and have fewer total payments to make. Being able to see how each option affects the overall money available for payments and the timing of becoming debt-free helps people make decisions.

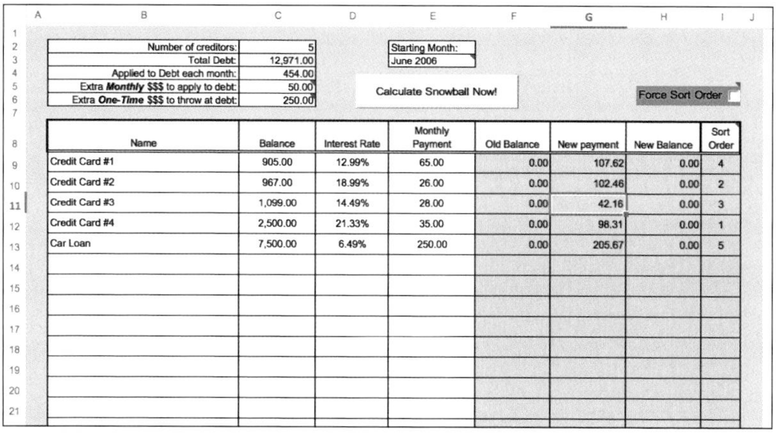

	Name	Balance	Interest Rate	Monthly Payment	Old Balance	New payment	New Balance	Sort Order
	Number of creditors:	5		Starting Month:				
	Total Debt:	12,971.00		June 2006				
	Applied to Debt each month:	454.00						
	Extra **Monthly** $$$ to apply to debt:	50.00		Calculate Snowball Now!				
	Extra **One-Time** $$$ to throw at debt:	250.00					Force Sort Order	
Credit Card #1		905.00	12.99%	65.00	0.00	107.62	0.00	4
Credit Card #2		967.00	18.99%	26.00	0.00	102.46	0.00	2
Credit Card #3		1,099.00	14.49%	28.00	0.00	42.16	0.00	3
Credit Card #4		2,500.00	21.33%	35.00	0.00	98.31	0.00	1
Car Loan		7,500.00	6.49%	250.00	0.00	205.67	0.00	5

FIGURE 4.12

Users can play around with different payment strategies in the snowball debt calculator to determine how they'd most like to tackle their debt.

Recharge the Batteries

People can make better choices when their physical and emotional needs are met. Being tired, hungry, or cranky makes it much harder to work on behavior change. That's why some behavior change apps will encourage users to indulge in self-care like eating regularly and well or sleeping consistently, even if the behavior in question is not specifically related to those activities.

There are also some apps that focus only on self-care in order to support other behavior change endeavors. Aloe Bud is an app dedicated to getting people to care for themselves so they're better prepared to tackle professional or creative projects (see Figure 4.13).

Work with Bias

People have a predictable set of cognitive biases that color their decisions. You can leverage those patterns to increase the likelihood that users make choices that get them closer to behavior change goals.

For example, you can use the anchoring bias to encourage more generous behavior. When New York City taxis began offering passengers the ability to pay by credit card instead of cash, the average tip amount jumped dramatically. It wasn't just that people's generosity was no longer limited by the contents of their wallet. Research found it had to do with the suggested tip amounts that displayed on the screen during the credit card transaction. The smallest suggested amount was larger than the at-that-time average tip. Even though people had the ability to type in any amount they wanted, the suggested amount triggered the anchoring bias and people used it as a guideline for their tip. Lyft, Uber, and other rideshare services have borrowed this design to boost their drivers' tips (see Figure 4.14).[2]

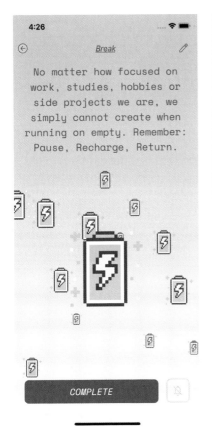

FIGURE 4.13
Aloe Bud reminds users who may see self-care as a distraction that it's actually an enabler of high performance.

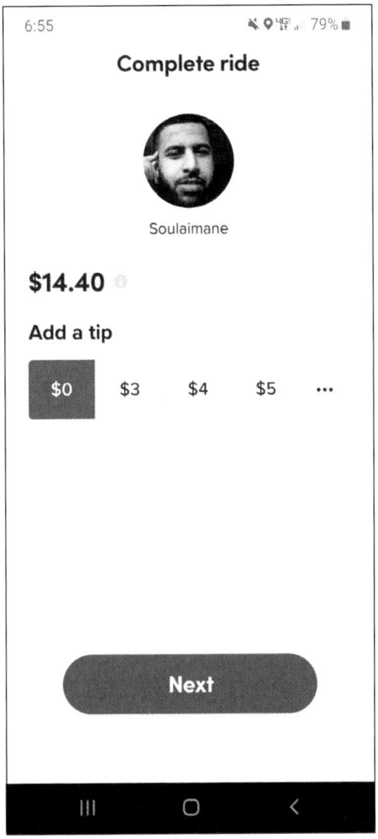

FIGURE 4.14
Rideshare services and electronic taxi payment systems anchor users' expectations of an adequate tip at a higher price point than is common among people tipping with cash.

2 Usually, it's easy to adjust tip amounts to be lower than the anchors, and gig workers like Lyft and Uber drivers are not well compensated, so I view this particular nudge in context as benign. That said, this is the sort of tactic that can be used to manipulate people and should be thoughtfully implemented.

A design technique that taps into both loss aversion and the sunk cost fallacy is the acknowledgment of "streaks," or patterns of positive behaviors over time. One way to reward a streak is with a badge on a user's account profile. If the user doesn't maintain the behavior, the badge could begin to fade or display a countdown to its disappearance, a technique known as "withering." Or a streak could be rewarded by displaying its duration (see Figure 4.15). This design makes it implicitly clear that the streak will be lost if the person skips the behavior. Both designs stir up users' negative emotions of losing something and remind them of the effort they've invested that will be "lost" if they don't maintain the streak.

FIGURE 4.15

After a whole year of weighing themselves daily, what user is going to casually break this streak?

A Spoonful of Sugar

Behavior change design is not about forcing a certain set of choices on someone. It's fair game to try to sweeten the deal by highlighting the positive aspects of behavior change–supporting options, as long as you aren't being deceptive or coercive to your users.

One tactic is to make the "right" choice more fun than the alternative. Volkswagen Sweden and DDB Stockholm worked together to see if they could use fun to get more people to take the stairs instead of the escalator in a Stockholm subway station. Their concept was a staircase painted like a piano keyboard that played music when people stepped on it. The piano stairs (see Figure 4.16) resulted in a 66% increase in people choosing the stairs.

One behavior a lot of people would like to see change is smokers tossing cigarette butts on the pavement instead of into a disposal bin. A UK company called *Ballot Bin* used the concept of fun choices to help accomplish this. They manufacture outdoor ashtrays that have two compartments and clear glass fronts so people can see how many butts have been tossed into each (see Figure 4.17). The ashtray can be customized with a poll question like "Do you use an Oxford comma?" and then the compartments are each labeled with an answer option—in this case, yes for people on the right side of history and no for the monsters. In areas where these Ballot Bins are installed, there's a 46% reduction in cigarette litter.

A digital program that combines fun with running is Zombies, Run! Zombies, Run! lets users choose a narrative

FIGURE 4.17
When people can use their cigarette butt to vote in a poll, they're less likely to toss it on the ground.

to listen to while they run that coaches them to escape a horde of imaginary zombies chasing them (see Figure 4.18). Instead of focusing on the running itself, Zombies, Run! users can put their attention toward the game. Working out is suddenly much more fun.

Run in the Real World.
Become a Hero in Another.

Only a few have survived the zombie epidemic. You are a Runner en-route to one of humanity's last remaining outposts. They need your help to gather supplies, rescue survivors, and defend their home.

And you have another mission — one they don't know about...

FIGURE 4.18
Zombies, Run! users are participating in a fitness program, but the primary focus is on escaping zombies by outrunning them.

So Good Looking

If you can't make the right choice "fun," you may at least be able to make it look more appealing. In digital design, you might see that "right" choices are presented in more eye-catching ways. Very often, this choice is whether or not to pay for an upgraded subscription to a program where users receive additional bells and whistles to support their behavior change.

Consider this example from Fabulous (Figure 4.19), where several elements of the design subtly nudge users toward the annual plan, which coincidentally is also the option more likely to generate revenue for its creators. Someone on a monthly plan may cancel after a month or two, whereas an annual plan member is locked in for twelve months. And, if the subscription is set to auto-renew, people on an annual plan are less likely to remember to cancel on time compared to people on a monthly plan. After all, the former group has twelve whole months to forget.

FIGURE 4.19

The annual subscription option is designed so that its savings over the monthly plan are emphasized three different times: in the header, in the feature list, and on the call-to-action button itself.

Similarly, Runkeeper orients people toward its paid plan over the free one in how the subscription options are designed (see Figure 4.20). The paid features are listed in the free account benefits column, but grayed out so that it's clear a user selecting free is missing out.

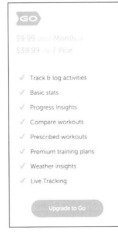

FIGURE 4.20

Runkeeper shows people the benefits of one choice over the other through design elements.

Products that guide users toward choices that result in a financial benefit for the company may seem like they're using design for evil. They (probably) aren't.

First, they aren't deceiving users into making a choice that they may not have otherwise made. These examples and similar ones from reputable companies clearly state what the user is buying and what the cancellation conditions are. Presumably, they honor these promises and make cancellation easy if the user requests it.

Second, if there's a good faith reason to encourage users to upgrade their subscription—so they can have a better experience and better results—then encouraging the upgrade isn't "evil." Of course, this assumes that the people making and marketing the product believe in it. Even better is if the people behind the app have done testing that suggests the upgrade is beneficial.

To gauge whether you're using this type of design nudge ethically, give yourself "the red face test."[3] Describe your design to someone who doesn't work with you, including how you encourage users down one decision path. If you feel any blush of embarrassment, or your audience reacts like you should, then revisit your design.

---

3 See also Nir Eyal's regret test: https://www.nirandfar.com/regret-test/

Give Permission to Be "Bad"

Since people are sensitive to cues that suggest what the "right" option might be, try to avoid using those sorts of cues in your design unless there's a valid reason to recommend a particular choice. You can reduce social desirability cues by:

- Emphasizing your users' privacy. In the example from Happify in the last chapter, users can opt into having extra privacy at the expense of receiving community encouragement. Offering them this option may facilitate more authentic progress through the program, which deals with the sensitive topic of emotional well-being.

- Reminding users that their choice is personal, and whichever paths they select are acceptable (which you know is true since you've curated their options to be good ones!).

- Letting users know that what works for someone else might not work for them; behavior change journeys need to be tailored to individual needs, preferences, and life contexts to be their most effective.

You can use microcopy within your product to reinforce that any choice the user makes is okay. Describe options in equally positive terms (or try for value-neutral language). Once a user has made a choice, acknowledge it positively ("Great! Good choice!"). Seize opportunities to be supportive of users' choices, whether or not they're the ones you'd have made yourself. In this example from Fitbit (Figure 4.21), users who bypass a specific fitness goal and say they're "just curious" receive positive feedback on their choice.

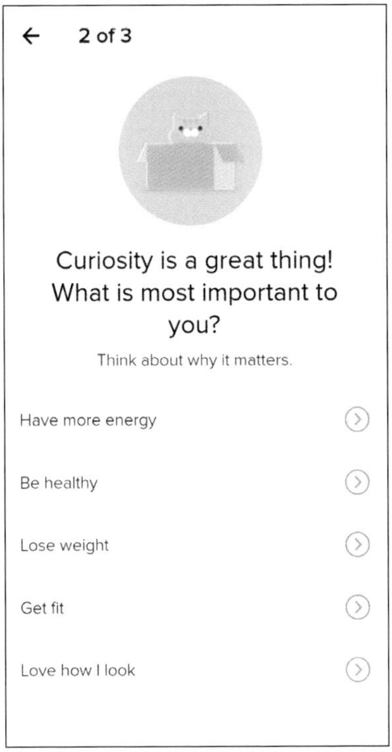

Curiosity is a great thing! What is most important to you?

Think about why it matters.

Have more energy ⊙

Be healthy ⊙

Lose weight ⊙

Get fit ⊙

Love how I look ⊙

FIGURE 4.21

"Just curious" may not be a behavior change goal, but Fitbit's designers recognized that people who choose it may just need some encouragement to stick around so they can decide what they want to do.

Choice Is Not a Weapon

Behavior change designers, perhaps more than any other type of design professionals, need to be on the side of their users. Effective behavior change design enables and empowers people to work toward meaningful goals. Effective behavior change designers help to simplify and structure good choices for their users so they can get there more quickly, without unnecessary angst.

So be on the lookout for choices that are potentially confusing, complex, or misleadingly presented, and use your design chops to make them more palatable for users. In the next chapter, you'll learn about how to identify some of the barriers that might make choices—and other ingredients of behavior change—more difficult for users so that you can design around them.

The Upshot: Keep It Simple

Choice can be hard for some people, but there's no reason for designers to make it harder than it has to be. Look for opportunities in your design to streamline decision-making for users. Whether it's curating options to help guide choice, presenting information so that it works with the way people naturally think, or highlighting the appeal of "good" options, try to reduce the brainwork required for users to choose.

With choice architecture, it's especially important to remember not to be evil. Give people the information they need to make an informed choice, including the drawbacks to their options. If you nudge people toward a particular choice, be sure that it's a benevolent one, and not designed to improve your profits or otherwise trick people into something they wouldn't willingly do. And never deliberately complicate a choice to dissuade users from going down a certain path. Keep it simple, and people will engage.

Speaking of easy choices, one of mine was asking Aline Holzwarth to offer her perspective for this chapter. When it comes to the topic of designing choice architectures, there are few people as knowledgeable as Aline. She applies theory to a wide range of behavior change challenges in her work. I asked her to share her thoughts on facilitating good user decision-making through design.

How do you think about choice architectures?

Friction is anything that gets in the way of a behavior, and fuel is anything that promotes that behavior. Humans are cognitive misers and tend to avoid using too much energy processing information. Because of that, even very small sources of friction can really get in the way, anything from an extra field on a website or having to sign a piece of paper and mail it in. These very small barriers can really deter someone's very good intentions, particularly when the thing they're trying to do has benefits far off in the future.

There are some cases where additional steps can actually serve as fuel. One thing we do in the Pattern Health app is give some people the choice of tracking their progress through a little Tamagotchi pet. This is adding another step. If they choose the pet, then they get to name the pet. That's also another step. However, instead of acting like friction, these steps are pleasurable and rewarding, so they actually act like fuel. And this lasts over time—now they are attached to this adorable little Tamagotchi. If they forget to take their medication, their little turtle is sad. They want to behave well in order to help their little animated pet. This is just one example of the malleability of friction and fuel, and it really highlights the importance of context and taking the user's perspective into account when we design environments.

How can you simplify choices?

First, we have to understand that every choice has an associated cost, and then weigh that cost against the benefit. When choices are unnecessary, we should simply eliminate them. As choice architects, we have the ability to decide which decisions are really important for people to make. Do we really need to ask for a ton of information during onboarding, or will that overwhelm people and make them give up? How can we prioritize the information that is truly necessary? If you think about an onboarding process, do you have to ask all the questions at once before someone actually gets into the app or can you phase it out? Breaking information into smaller chunks and delivering it over time is easier on the user—and that's what we really care about.

Two, smart defaults can be incredibly helpful. A default is when you set what something will be if the user takes no action, but still give the user the freedom to opt out or change their decision. So if you need to send reminders for someone to take their evening medications before bed, you can set a default routine where you say, "This is what the average schedule is." Then you can allow users to change that if it doesn't work for them. There are people who don't live within this average schedule and that's fine. They have the ability to change it, but they don't have to put in all that cognitive effort if the average routine works just fine for them.

How can you break down behaviors for users?

One thing that's very tempting to do that is not very effective is focusing on these really overwhelming goals like, "I want to have more energy." These things are so abstract that we have difficulty figuring out in the moment what the steps are that we actually have to go through. We can guide people through those logical steps that will ultimately land them with their desired outcome in the end, focusing on more specific concrete tasks that they have control over. You don't have control over how much energy you have. You do have control over how much caffeine you have, or what time you go to bed. Those are the kinds of things that we can get people to focus on, the steps that lead to the ultimate outcomes.

We can give people choices in ways that are empowering but not over-whelming. We did a weight tracking study with heart failure patients where we asked them to set an implementation intention for when to step on the scale. They can choose before or after dinner. Dinner then becomes a cue that they need to step on the scale. This preserves the feeling of control by letting patients choose when they want to take the measurement, and as a nice little side effect, it has this other benefit of reminding them to actually do it.

 *Aline Holzwarth wears three professional hats as the Head of Behavioral Science at Pattern Health, principal at the Center for Advanced Hindsight at Duke University, and president and cofounder of the Behavior Shop. She's also a prolific writer whose work can be found on Medium, the Center for Advanced Hindsight website, and in the* Wall Street Journal *and* Scientific American.

CHAPTER 5

Something in the Way

Diagnosing Ability Blockers

The point of behavior change design is to facilitate the performance of target behaviors. With every design decision you make, you should return to the core question: "What do I want my users to *do*?" (Hint: It's the target behavior.) Whether it's reading content, making a purchase, or sharing data, every digital product has something that it asks its users to do. *Your job as a designer is to make those behaviors as easy as possible for your users.*

A major factor in whether or not someone performs a behavior is whether or not they can. That may seem so obvious that it doesn't need to be said, but designers forget it all the time. When you have expertise in how to do something the right way, you might assume that other people have a similar level of knowledge. Experts can forget what it was like to be a beginner. Many times when a product that's supposed to help people change behaviors fails, it's because the designers didn't understand and address the reasons why that particular behavior can be hard.

On the other hand, when you know what stops people from doing the desired behavior, you can design to overcome those barriers. You can provide tools or product features that make a behavior easier; you can figure out ways for people to avoid the barrier altogether; or you can equip people so that the barrier is no longer an issue. If you can remove the things that stop people from doing a behavior, the chances that they will do it become much, much higher.

But first, you've got to figure out what those things are. This chapter will introduce a framework to help you do that.

What Blocks Ability?

The tools in this chapter are focused on understanding what limits people's ability. I use the term "ability blocker" to refer to anything that constrains ability.

The first step in designing for ability is to figure out what ability blockers are in play for your users. Think about your target audience, what you're asking them to do, and the context in which they'll be interacting with your product. What might get in the way of your users performing the behaviors you want them to do? These could be factors related to the product itself, the environment it's used in, or something about the user. Some common ability blockers include the following:

- **Knowledge:** Is this a confusing, complicated, or new area for the people using your product?

- **Skill:** Does performing the behavior require talent? Do people need to practice to do it correctly?

- **Time:** Do people need to dedicate a chunk of time to the behavior? Anything that takes more than a few seconds to do might be hard for people who are busy.

- **Focus:** Are people using your product in a noisy or distracting environment? Does the design of the product make the objectives clear, or are other features within the product itself drawing attention away from the main parts?

- **Mood:** Are your users experiencing stress or anxiety? Low levels of these moods can enhance ability, but too much will prevent people from performing behaviors well.

- **Tools and resources:** Does the person need to have special equipment or specific items to perform the behavior? Do they have to spend money to do the behavior?

- **Motivation:** Does the behavior align with the person's goals? Is it something they *want* to do? Does it outweigh other important goals, like putting food on the table?

As you can tell from this initial list, people can have multiple ability blockers affecting a single behavior. For example, if someone doesn't have the right equipment to perform a behavior, it may be the case that they also don't have the money or the time to get that equipment. The closer you can get to the root cause of why the person isn't performing the behavior, the better you can address it.

There are four basic ways you can identify ability blockers for your users and target behaviors: using common sense, conducting informal research, doing a literature review, and conducting formal insights research.

Use Common Sense

Often, you can generate a good starter list of things that might reduce a person's ability to behave the way you'd like them to by just using common sense. Think about your own experiences with the target behaviors and jot down the ability blockers you've encountered. Brainstorm about what might affect other people.

Conduct Informal Research

Informal research can be a great way to learn about ability blockers from real people quickly and at a low cost. Reach out to your networks to find people who have experiences with the target behaviors you're designing toward or solicit introductions to those people. Look on social networks for well-trafficked hashtags related to the behavior and read what people are posting. Then reach out to anyone with interesting comments to see if they're willing to talk. Once you've identified your research sample, ask them what was hard about the target behavior and what helped them do it.

You can also spend time around people who are already doing your target behaviors and pay attention. What do you see? If you're working on a program to help people get physically active, what are exercisers in your community doing? If your product is a mobile app, watch people using their phones in different contexts. Do you see anything that might affect how people use your app?

Do a Literature Review

Chances are, someone has done research on your target behavior. Look up published studies on your target behavior using Google Scholar or an academic database. A literature review will give you a head start on seeing what solution types have and have not worked in the past.

Conduct Formal Insights Research

Formal, systematic research may help you uncover less obvious ability blockers. If you have the budget and opportunity to include primary insights research in your product development process, then do it. Later in this chapter, I'll share a process for organizing the research you do to better detect ability blockers.

Once you actually build a digital product, you can take the list of possible ability blockers generated by your research and try to see which ones apply to any given user. The screenshot in Figure 5.1 shows how HelloMind asks users to describe what types of motivational problems they're experiencing.

As you learned from Vic Strecher in Chapter 3, "It's My Life," not everyone has good insight into their own motivation at the outset of a behavior change process. Be prepared to test different ways to ask about blockers and motivation to see what works for your users. These may include breaking the questions down into more manageable chunks ("What's the wallpaper on your smartphone?") or coming in from an entirely different angle ("Which one of these characters is most like you?").

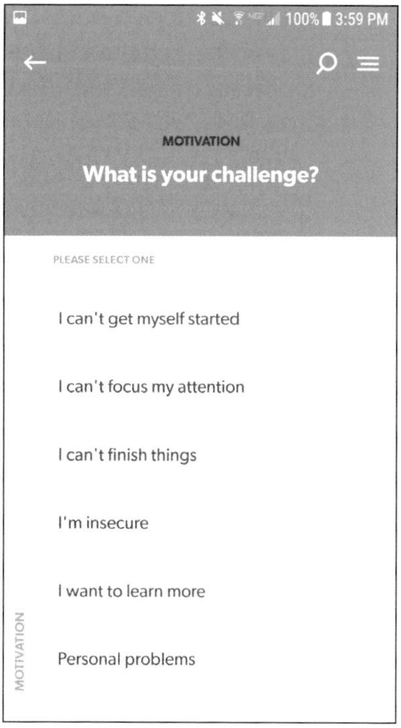

What Can Insights Research Tell You?

Here's one example of the richer detail you can get from formal insights research into ability blockers. Why do construction workers take risks on the job? My team was hired by a risk management firm that wanted to understand why construction workers sometimes disobey safety protocols. That knowledge would help them design a training program to reduce the number of serious injuries and fatalities (SIFs) on construction sites.

We conducted observations at three construction sites. Before we went on-site, we reviewed existing research about construction safety so that we knew what types of behaviors to keep an eye out for. As we walked around, we saw a few people standing on the top rung of a ladder without anything to hold on to and a number of workers who hadn't worn their personal protection equipment (PPE) right. These behaviors could potentially lead to SIFs.

Then we interviewed workers and supervisors. We heard that management cared about safety, but there was also pressure to get work done quickly. This meant that workers might use the equipment close at hand instead of spending 10 minutes to get the right items from a trailer on the other side of the job site. It was a lot faster to stand on the top rung of the ladder they had rather than stopping to get a taller ladder.

FIGURE 5.1

HelloMind's designers have narrowed down the universe of ability blockers related to motivation that their users might experience. Within the app, they ask users to pick which ones are most relevant for them.

Many of the workers didn't want to admit they'd done anything unsafe. In some cases, they may not have even realized they did anything unsafe. Because we'd been able to directly observe some unsafe behaviors, we were able to ask questions about them without our interviewees feeling personally blamed. As they became more comfortable during their interviews, many of the workers eventually shared stories about their own safety shortcuts. We needed both the observations and the interviews to piece together the complete story of why people might stand on top of a ladder.

The Behaviour Change Wheel

After you've decided on how you'll identify ability blockers, through research of some kind or a common-sense cataloguing, you'll want to have a framework to collect and organize information. A helpful tool for categorizing ability blockers (and later finding the best types of solutions for them) is the Behaviour Change Wheel.[1]

The Behaviour Change Wheel hinges on a system called COM-B. The COM-B model simply says that in order for a Behavior (B) to occur, people must have enough

- Capability (C)

- Opportunity (O)

- Motivation (M)

to perform it. See Figure 5.2, which shows the COM-B model. Notice that capability and opportunity can influence motivation. People may feel more interested in a behavior when their personal skills or the environment makes it easier to do.

Each type of ability blocker is best addressed by specific solutions. After you've identified whether your users struggle with capability, motivation, or opportunity, the Behaviour Change Wheel helps you pick the right types of solutions to help them. Although the system is called a wheel, I think of it more as a decision tree. In Chapter 6, "Fix You," you'll see a grid that links from barrier types to solution types.

---

1 The Behaviour Change Wheel was developed by researchers at University College London based on a systematic literature review of more than 1200 studies and 19 different systems of behavior change. The researchers identified patterns and put them into a framework. It's open source and can be used as the basis for your own research protocols and tools.

You'll also want to determine which blockers are the most important ones to address with your design. Later in this chapter in "Organizing Your Research to Detect Ability Blockers," I'll share one way to prioritize which blockers to tackle first.

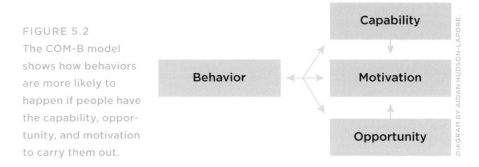

DIAGRAM BY AIDAN HUDSON-LAPORE.

FIGURE 5.2
The COM-B model shows how behaviors are more likely to happen if people have the capability, opportunity, and motivation to carry them out.

You'll need to understand capability, opportunity, and motivation and how to recognize ability blockers in each category. Evidence for these types of blockers in your users could come from talking to people, observing them, or third-hand accounts, so the descriptions include clues from a mix of data types. Because many behavior change products ask people to make changes in their physical environments outside of the product, your research should consider capability, opportunity, and motivation both on and beyond the screen.

Note that the word "motivation" within the COM-B framework is used in a broader way than I've used it throughout this book. For purposes of explaining this framework, I've used its terminology. Outside of COM–B-specific content, I use motivation to mean "desire with velocity." Within COM-B, motivation includes that definition under the subheading of reflective motivation, while also including the concept of automatic motivation to cover belief systems that influence behavior.

Body and Brain

The first category of ability blockers is called *capability*. It refers to the physical and psychological abilities of a person to complete a behavior.

Physical capability is pretty much what it sounds like—does the person have the physical ability to perform the required behavior? For digital products, think about perceptual abilities such as eyesight and hearing, as well as fine motor control for mousing and clicking. Basic usability best practices will eliminate a number of capability

ability blockers at the outset. Your colleagues who focus on usability are important partners in behavior change design.

Accessibility in digital design refers to making your products usable to people with disabilities. There are several comprehensive resources for understanding accessibility best practices, including W3C and Usability.gov, and many wonderful books that go deep on accessibility for different types of products and industries. Use these resources!

There are many reasons to incorporate accessibility best practices into the design of any product you work on. Consider: Why wouldn't you take the opportunity to pre-emptively avoid 15%+ of potential users dropping off because they had trouble using your product? The World Health Organization estimates that one billion people worldwide have a disability. That's 15% of the population.[2] Beyond that, not everyone who benefits from accessibility guidelines is technically disabled. People whose vision has worsened with age, for example, may benefit from some of the font and color contrast guidelines associated with accessibility.

Accessibility is not just about avoiding negative issues. It's also an avenue toward new ways of thinking about tackling challenges. Case in point: Microsoft's Adaptive Controller for Xbox won Fast Company's 2019 Innovation by Design product of the year award.

Psychological capability includes knowledge and skills.[3] In order to perform a target behavior, a person needs to understand how to do it and all of its component parts, *and* have the mental skill set to carry it all out. Let's take asking someone to "eat a healthier diet." Psychological capability to do this would include:

- Understanding what a "healthier diet" means

- Being able to plan balanced meals and buy the right ingredients

---

2 In January 2019, the Ninth Circuit Court of Appeals in the United States ruled that the Americans with Disabilities Act (ADA) applies to websites as well as brick-and-mortar businesses. This opens the possibility that future digital products may be legally required to incorporate at least some accessibility standards.

3 A related concept is Don Norman's mistakes versus slips. A *mistake* is likely to be a psychological capability error; people don't understand the right actions to take. A *slip* may be caused by something in the environment, like the toothpaste being stored where the Icy Hot normally goes, or distracting noises that make it hard to answer questions correctly.

- Knowing some recipes to cook or where to find them

- Having the ability to chop, peel, and dice the ingredients

- Finding time in the day to cook the meal

If the person is missing any of those bits of knowledge or skills, their ability to get the behavior done successfully will be limited.

When you're investigating whether there are capability issues blocking someone's ability to perform a target behavior, this is some of the evidence to seek.

Can the person complete the behavior start to finish correctly? Watch to see if the person makes mistakes, and what type. Or, if the person asks a lot of questions about how to finish the behavior or looks up additional information, that tells you they don't feel confident in their knowledge. Pay attention to whether there are these sorts of stutters and what they are.

Some products, like Grasshopper, ask users to provide information about what they already know (see Figure 5.3) to help quickly figure out and address any knowledge gaps. Other products may be able to segment their users and make grounded assumptions about their abilities based on that segmentation. You might imagine a coding app that automatically places users into a more advanced track if it detects specific other apps on their devices, for example.

Can the person complete a complex set of behaviors that add up to a big behavior in sequence, or do they need to stop at points along the way? An example might be needing to consult paperwork at multiple points while going through an onboarding process because the product is asking for unfamiliar information. One way to think about this aspect of capability is *stamina;* whether

Is this your first time coding?

Yes, I'm new to coding →

← No, I've coded before →

||| ◯ ‹

FIGURE 5.3
Grasshopper asks users to describe their experience writing code during onboarding, and provides more basic introductory lessons for people who need them.

the person has developed the resources needed to keep at a behavior over a long period of time.

What tools, if any, does the person need to do a behavior? Look for extra tools or assistive devices that a person is already using or could benefit from adopting in order to do the behavior. For example, a sleep and relaxation app might offer soothing nighttime meditation audio, but without comfortable headphones to avoid disturbing a snoozing partner, the meditation becomes difficult to do as intended.

Can the person explain how to do the behavior to someone else? Usually, being able to teach another person means that someone has gone beyond beginner status. Listening to their explanation can also provide insights into what they understand, where they still struggle, and where your design can make things easier. Usability researchers frequently uncover capability deficits when they ask research participants to verbally describe how they're using a product; the descriptions reveal gaps in understanding and areas where the design needs improvement.

How well can the person coordinate with others to complete the behavior? If the person needs to work as part of a team to accomplish the behavior, it's important that they're able to communicate effectively and coordinate activities. This might mean something like coordinating with a spouse so that one person buys the right groceries for the other one to cook a healthy dinner.

Can the person self-regulate when there are distractions or something goes wrong? If there are interruptions in the environment, like ambient noises or other people, it may make it harder to perform the behavior. People who are more skilled can typically adjust so they can complete the behavior anyway. The ability to calibrate behavior based on environmental feedback is called *self-regulation*, and it can help people stay on track, especially with newer behavior changes that aren't yet habits or routines. Self-regulation can be especially important for examples like the construction worker safety behaviors, since worksites tend to have a lot of people working on them while heavy machinery moves back and forth—not to mention all the noise.

You may notice that capability factors tend to be related to something about the person, whether it's their physical condition or a mental characteristic. Keeping that in mind can help you sort through whether an ability blocker belongs to the capability category or not.

A World of Possibility

Opportunity is about the environment in which a person might perform a behavior. It includes physical opportunity and social opportunity. *Physical opportunity* has to do with whether a person's surroundings help or hinder them in the behavior. For example, beginner exercise programs often emphasize walking as a gentle and safe way to begin physical activity. But some people live in places where there aren't well-lit sidewalks where they can stay safe from cars while they walk. Not accounting for that physical opportunity barrier leaves people who live in those areas unable to really benefit from an exercise program that centers on walking.

Social opportunity has to do with the other people who affect the behavior. It includes things like direct peer support, but also what people learn from observing others around them. People are heavily influenced by their social contexts, whether it's the larger culture they live in, or the social circles they're part of. Research has shown that people's social circles influence them much more than you might guess. It makes sense that you're much more likely to quit smoking if your spouse or best friend gives up the habit. But network effects are so strong that you're also much more likely to quit smoking if *friends of friends* do. Understanding the social environment people live in is critical to designing for ability.

Here are questions to determine if there is an opportunity issue at play:

- **How does the landscape where the behavior is performed make it easier or harder to do?** To understand physical opportunity, investigate the built environment. Pay attention to how the environment influences people's ability to do the behavior.

- **Are the materials or tools required for the behavior available?** Some behaviors require tools or materials to do. If they're not easy to come by, then there's a physical opportunity blocker. For example, if you're asking someone to strength train, are there weights or some other appropriate object they can lift nearby?

- **What kinds of time or schedule constraints affect the behavior?** If you're asking someone to go through a complicated medication regimen first thing in the morning when the kids need to get ready for school, they're likely to have some physical opportunity issues.

- **What are the social norms the person is most often exposed to?** Try to understand what's "normal" among the person's friends, family, or coworkers. Gather information about how people in those groups react when someone is different. Are they supportive? Openly critical? Most people feel uncomfortable bucking norms too much.

- **What behaviors are people who have some kind of authority doing?** People often model their behaviors on what they see others do, especially if it's someone who has position power (like a boss or supervisor) or whom they admire (like a celebrity or the most popular kid at school). The effect is amplified if others are also modeling their behaviors on the same person or people. As they say, everybody's doing it.

- **What other groups or teams is the person exposed to that might influence their behavior?** This one comes into play often with workplace behavior change. Look at how other groups make it easier or harder to do the target behavior. For example, it can be hard to get doctors to change a behavior if it doesn't mesh with the routines their medical staff and patients follow.

- **What types of explicit social support (or nonsupport) is available?** Sometimes, friends or others encourage the new behavior and participate. Other times, they create situations where the new behavior is harder to do, like sabotaging a friend's health efforts by organizing a night out at a bowling alley where everyone smokes and the only food options are fried.

A good rule of thumb is that if an ability blocker exists outside of the person who is trying to change a behavior, chances are it falls into the opportunity category.

NOTE DON'T FORGET ROVER!

When I was working on a cognitive behavioral therapy–based intervention to help people adopt good sleep habits, we first looked at the ability blockers that stopped people from sleeping well. Many of those habits were expected from looking at existing research, but one kept coming up that was not on our radars: pets.

Pets can both block and boost people's ability to sleep well. They can be blockers if they disrupt a person's sleep by coming in and out of the bedroom, making noise, or crowding the bed. They can be boosters if they're quiet and snuggly. Dogs can also boost people's ability to sleep because they require being walked several times a day. Well-timed exercise can help people fall and stay asleep, so walking Rover at the right time of day may lead to higher quality sleep at night.

Armed with this information, we asked people at the outset of the program to tell us about their pets, including names. Then we were able to personalize coaching about how to stop pets from becoming sleep blockers (mainly by keeping them out of the bedroom!) and how to take advantage of the sleep boost they could offer. For at least some of our users, coaching around this specific blocker made a difference.

Where There's a Will

Motivation is the most "psychological" of the three categories of factors that influence people's ability to do a behavior. Remember, when it comes to COM-B, the word "motivation" has a slightly more focused meaning than I've used so far. It refers to two subcategories: reflective and automatic motivation.

- **Reflective motivation** is pretty close to what most people think of when they hear the word "motivation" and the way the term is used in self-determination theory. It includes goals and values that people explicitly claim for themselves. When someone says, "I want to lose 10 pounds," that's an example of reflective motivation. Another example might be someone who tells you, "I want to set a good example for my kids." While that statement may not be tied to one specific behavior, it sets some expectations around what behaviors are desirable and which ones are not.

- **Automatic motivation** is a little trickier to understand. It includes all of the psychological stuff that people may not even really understand about themselves. For example, there are cultural expectations around behaviors that influence what types of things people think are acceptable to try. And there are also emotions and desires, which may not align with logical thinking but drive behavior anyway.

Assessing motivational barriers can be the trickiest of the three components of COM-B. That's because motivation, at least in the COM-B sense, contains a hodgepodge of factors that are hard to observe. And some of the items that fall under motivation may not be easy for people to tell you about if you ask them.

For example, a lot of people might behave a certain way because of role identity: "This is how a father acts." Unpacking that is a little like a therapy session that not everyone is able or willing to participate in. Many times, behavior change designers are left to infer what really falls into the motivational bucket based on what they can see and what they hear under the surface in talking to people.[4] It's a prime opportunity to combine interviews and observations for better data. And have some faith; like so many types of UX research, COM-B–rooted inquiries can feel overwhelming at first, but you *will* see themes as you talk to more people.

- **What are the person's goals?** People may have thought about their goals and be able to articulate them. If so, great! If they can't, ask how they would describe the future if they were to be successful at behavior change. Their answer will give some insight.

- **Which goals are most important to them?** It's common to have multiple goals, but sometimes they compete with one another. For example, often people have personal health, fitness, and finance goals, but their goal to be a good parent takes precedence. So it becomes pizza and Disney trips instead of salad and 401(k) contributions. Understanding goal priorities helps you design in a way that takes account of what matters most while still making progress toward less critical goals.

- **How do people understand their roles?** This may be easier to assess in group scenarios like a workplace, where people have a job title. Although it might be harder to articulate, most people also have a personal meaning around what it means to be a friend, a partner, a family member, etc. Those role definitions can limit or expand the set of behaviors people feel comfortable doing while playing their roles.

- **Does the person believe that the target behavior will help with any of their goals?** People aren't generally as interested in doing things they don't think will help them be better off. This blocker comes up often when people have failed at a behavior change in the past. "It never worked before, why will it now?"

---

4 To learn a method that helps you listen for the meaning beneath the words, read *Practical Empathy: For Collaboration and Creativity in Your Work* by Indi Young.

- **What types of rewards are there for doing the behavior?** These may be financial incentives (common in employer-sponsored health plans, for example, where people may pay a lower cost if they complete certain activities). Or they could be material incentives, like booking a luxury vacation as a reward for finishing a tough work project. A paycheck is a very common example of a reward people receive that causes them to prioritize a certain set of behaviors.

- **What types of punishments or negative consequences exist related to the behavior?** For example, does the health insurance premium cost go up if the person doesn't complete those wellness activities?[5] Some programs, like stickK, have users wager money on their goals (see Figure 5.4). If the person doesn't succeed at their task, they lose the money. An added twist is that users can additionally designate any lost funds toward an "anti-charity," a cause they don't personally support, to make losing the money that much more painful.

FIGURE 5.4

stickK users can wager money on their behavior change efforts. They further designate where any lost funds are sent, which may add fuel to the follow-through fire.

Automatic motivation is by definition not easily accessible to people. Again, that's why listening for the story behind the story is so critical. Comparing what people *say* to what you see them *do* can

---

5 In Chapter 4, you learned about loss aversion, a cognitive bias in which people are more sensitive to losing something they already have than the promise of gaining something new, even if the bottom line is the same in both situations. This is one way people leverage that bias in behavior change design. Many programs have incentive structures that offer penalties (taking away an inexpensive option) instead of rewards. This can be effective, but avoid overusing it. Whether it's a health plan or a low-cost airline, people tend to feel a lot less happy if they think they're being nickel-and-dimed for every behavior.

also provide clues as to what's motivating them. Some things to look for include:

- **What kinds of consequences are associated with doing the behavior or not?** The social and emotional outcomes of a behavior strongly influence people's interest in doing it. Behaviors that garner respect are more appealing. Avoiding behaviors can also have positive consequences. If someone believes they're likely to fail, they may prefer to avoid trying.

- **How does doing the behavior make people feel?** A different type of consequence is whether the behavior is pleasant to do. This could include mood (Is it stressful? Scary?) or a physical experience. As a rule of thumb, people seek out experiences that feel good and avoid experiences that feel bad.

- **What context will people be in when they interact with your product?** Understanding the physical and emotional context around a target behavior will help you determine if that behavior is rich with feeling for someone. For example, someone who is waiting for a family member in an emergency room is going to bring a very different mindset and emotional stance to your product than someone relaxing on a couch at home with their smartphone.[6]

- **What type of cultural understanding does the person have about the target behavior?** This is the big-C Culture of the context in which a person was born, raised, and lives. If someone has been raised in a culture where a certain type of behavior is very normal or is never done, that colors their willingness to try the behavior themselves. In some ways, it's easier to tackle cultural effects on automatic motivation when working internationally because the contrast makes them easier for designers to see.[7] But every country has many microcultures that deeply influence the people who grow up in them. It may take more work to see them.

---

6 *Design for Real Life* by Eric Meyer and Sara Wachter-Boettcher offers a primer on designing for people in extremely difficult situations. It's highly relevant for anyone designing for people who may be in a hospital, health system, or other medical environment (among others).

7 But please note that it is critical to include the people you're designing for in your process. Once you notice cultural differences, explore them with the help of a local guide.

It's worth noting that the COM-B model defines ability a little bit differently than I have here. I'm using a broad definition of ability that covers capability, opportunity, *and* motivation. Don't get too hung up on the terminology—the underlying concepts are the same. Anything that's getting in the way of people performing the desired behaviors is something you want to consider in your design.

Organizing Your Research to Detect Ability Blockers

A great way to use COM-B and the Behaviour Change Wheel in the behavior change design process is to plan and conduct research (what I called the *diagnosis phase*). Assuming that you plan to use it as a way to design for ability, it's a good idea to set up the basic structure of your work to follow the Behaviour Change Wheel from the beginning to eliminate rework and ensure that you get all of the inputs you need. A bit more time spent in planning will save a lot of time in the analysis and interpretation.

Write Interview Questions and Observation Checklists

If you are conducting interviews (even informal ones with friends and family), you'll want to have a list of questions to ask. It doesn't have to be as official as a moderator's guide, but, at the very least, have your main questions written out in an order that makes sense for a conversation.

The descriptions of the types of ability blockers are peppered with questions on purpose. Those broad starting questions can be the basis of your moderator guide for interviews. Take that list of generic questions about capability, opportunity, and motivation and modify them for the specific project you're working on. You might put more questions around an ability blocker type you have reason to believe is especially important for your target behaviors—for example, if you've already done a lit review and seen that something keeps cropping up as an issue. If you know an ability blocker is irrelevant for your target behavior, don't spend much, if any, time exploring it.

If you're doing observational research, create a checklist of things to look for. You can source items by seeing what prior research turned up, or do some quick-and-dirty pre-research by talking to people who have experience in your topic area. Then when you're on-site, you can keep an eye peeled for the relevant evidence. A checklist also makes note-taking easier since you can just mark the relevant items, with short phrases to provide detail. (If you have permission, take lots of photos!)

Create a Data Grid

An easy way to organize the data you gather about ability blockers is by creating a grid, either by drawing a table or creating a spreadsheet. The rows of the grid should be blocker types. You can make these as broad as capability, opportunity, and motivation, but usually I get to a more detailed level with three to four flavors of ability blockers under each major category. The columns should include evidence for this ability blocker and space to rate how critical this ability blocker is. Sometimes I also include a column for enablers, when a particular category of ability blocker is working in people's favor. For example, if the intervention is to get people walking outdoors and they live in a community with a well-lit, scenic downtown, take note of that so you can leverage it in your design. In this case, there's a physical opportunity booster for the behavior of walking.

Create a grid like the one in Figure 5.5 to organize your evidence for ability blockers. The grid in Figure 5.5 focuses on an intervention to get people to work out more often. In a real project, you'd probably have more information in the grid, but you can see in the sample how you can add additional rows for each flavor of blocker. The evidence column includes data from your research participants (abbreviated by number), which can be either quotes or notes. If you like to color code, you can make blockers one color (I used red), and facilitators or boosters another color (green, in this example).

You'll add to this grid later to make decisions about which blockers to prioritize solving for in your design. You can also include a column to identify the solution for that ability blocker, to facilitate the next stage of the design process.

How to Handle Overlapping Data

You also might notice that there are lots of blockers that seem to fit multiple COM-B categories. In that case, tag them for all of the categories you think they belong to. Then put the information in the grid next to the category you think best describes it (or if you're truly

ABILITY BLOCKERS NOTES GRID: EXAMPLE

COM-B Category	Subcategory	Evidence for Existence of Blocker
Physical Capability	N/A	
Psychological Capability	Knowledge	P2: "I get confused by all the steps I'm supposed to do. What comes first? I always have to look back at the app to check. And then it takes so long to do the workout." P4, 6, 7, 13 all use some kind of cheat sheet to get through workouts
	Self-regulation	P5: "I can do the yoga sequence really well if I've got the energy and a quiet space to focus, but as soon as my kids come home it's game over."
Physical Opportunity	Equipment	P7: "I don't even know where to buy this stuff! And where exactly am I supposed to store it in my studio apartment?"
Social Opportunity	Social support	P14: "I know my husband wants me to be happy, but he has a really hard time when I don't want to sit around with him watching TV anymore. He makes me feel bad about abandoning him and the things we used to enjoy together." P11: "It turns out my good friend also wanted to sign up but was afraid to do it alone. So now we're gym buddies. I know if I skip, I'll be disappointing her, so I go."
Automatic Motivation	Behavioral Consequences	P4: "The last time I went to the gym, I couldn't walk right for three days afterwards. Everyone was staring at me limping around and my legs would cramp up. I can't live like that."
	Emotional Response	P14: "I just feel like crap when my husband gives me that look when I tell him I'm going to do a yoga video in the bedroom instead of watching Netflix with him."
Reflective Motivation	Goals	P2: "My high school reunion is next year, and I keep thinking how nice it would be to feel confident when I walk in the room."
	Priorities	P2: "I figure I should work out four times a week if I want to hit my goal. But then I also work so much, and I barely see my kids as it is, and I have to put food on the table and I'd like to be there for bedtime. So realistically I maybe make it once, twice a week."

FIGURE 5.5

A grid like this one is a good way to organize your data and observations against the types of ability blockers to see what gets in the way for your target audience.

torn, toss a coin). For the other categories that work, make a note that cross-references the first category. Don't stress out too much about making the perfect choice. Usually, if a blocker is hard to categorize, it's because it truly can be explained in multiple ways.

A really cool thing about using the Wheel is that these sorts of overlapping categories tend to sort themselves out when it comes to choosing a solution. Very often, the different category options for a single blocker share one or more common solution types. Seeing the same type of solution come up for the blocker across all of its potential categories should give you some confidence that you've achieved triangulation around an approach.

In the sample grid in Figure 5.5, P14's ability blocker related to her husband's lack of support is referenced in both social opportunity and automatic motivation. You can imagine that if there's a way to help her husband get on board with her new fitness routine, it might address both categories.

The next chapter focuses on the solutions that the Behaviour Change Wheel connects to for each category of ability blocker.

The Upshot: Be a Detective

There are many reasons why people may not perform a target behavior or find it difficult to do. In order for your behavior change design to achieve its objective, you need to understand what's blocking people from the target behavior. You should investigate potential behavioral blockers early in your project, whether it's through formal project research or more informal methods.

The COM-B model provides a helpful tool to organize your research into categories of blockers. Think about whether the blockers you find are related to capability, opportunity, or motivation. Having this categorization will allow you to use the Behaviour Change Wheel to pick intervention features that are more likely to actually work.

Steve Portigal knows research. Although Steve doesn't specialize in behavior change, many of his projects ultimately focus on understanding why people behave the way they do so that designers can help them do something differently. Perhaps more importantly, Steve is a master at discovering the unexpected insights that differentiate successful projects from the rest. As Steve put it during our conversation, "That's why you do research: To find what's different from your expectations." I asked Steve to share some expert tips for using research to learn about people's barriers to behavior change.

How do you understand what's driving a behavior?

You can *ask* people what they're going to do, but you shouldn't necessarily take them at their word. It might be a great question because their answer reveals their mental model, but that's not actually what they're really going to do. If you talk about it for an hour, get stories, and ask really good follow-up questions that get them to focus on details, people become more engaged. You help them through the course of a good interview to be more insightful about their own behaviors than they might otherwise have been.

Think about having a conversation with people where they talk about the process of the behavior. They talk about what it is that they do, and you might ask a clarifying question about some aspect you're trying to understand. But even having people say what it is they do without explaining tends to reveal where their mental models may differ from yours. Your thoughts as the maker of something, the designer, are often different from how someone else describes how they engage in this behavior. It behooves us to listen closely to what is really being said, because that gives us perspective on how the person is constructing their approach to that behavior. Use your spider sense. It's a feeling that says, "Wait. Something is not quite right or it's different than what I expected." Be able to pause in those moments.

Why do you have participants prepare materials before an interview?

It's less about the reams of data that we gather and more about the change that data collection instills in our participant. The data collection activity can be very low key. We can ask people to collect a sample of artifacts. For instance, I've had people save things like junk mail that comes in, or wine bottles that they had finished (for a wine packaging study). If I can go to somebody's house and they have thirty bottles of wine versus four, then there are more samples to look at.

What I find exciting is if you're having people (for example) log every banking interaction in a spreadsheet, and then maybe write a sentence about it, they're now more mindful of that behavior. This puts them in a state where they have the ability to reflect. People start to spontaneously tell me things like, "You know, since I've been collecting these things, I've noticed this..." They seem to be more insightful.

How can you uncover barriers to new behaviors?

We were working with an organization that was on the cutting edge of a change in how people go about an everyday activity, but more broadly this hadn't happened yet. The organization's question was, "When people do make this change, what barriers do they have to overcome?" We had an actual product. And so what we did was find people that were likely users, and induced them willingly to step over the line to making the behavior change. We gave them the product and workbooks that had daily and weekly assignments. We artificially created new users, and then asked them to document what happened.

That gave us a lot of stories to figure out the ideal design of the experience to support people through making this change. It helped us understand what life's like through that gate, so that we could better design the entryway.

What if an interview goes off track?

Things can go awry or feel like they're not where you want them to be. Part of the work of the researchers is to figure how to get back on track. I think it's an interesting exercise to just persist. Maybe you can just keep going, and there's something there. If there isn't, it's still good practice.

And then, you don't always know in the moment what is going to be insightful. There are lots of stories about people coming out of interviews being frustrated and feeling that was not a good use of their time, and then finding themselves going back to what that person said later on. It's all about not giving up too quickly. Maybe the interview will play out. If it doesn't, what have you lost?

 *As a consultant, **Steve Portigal** conducts user research to inform product development and strategic roadmaps. Steve also coaches organizations to level up their own research capabilities. Steve is the author of* Interviewing Users: How to Uncover Compelling Insights *and* Doorbells, Danger, and Dead Batteries: User Research War Stories, *and hosts the* Dollars to Donuts *podcast featuring other leaders in user research.*

CHAPTER 6

Fix You

Solving Ability Blockers

So you've figured out what's stopping people from performing the target behavior or behaviors? That's great! But unless you design something into your product to help them overcome those ability blockers, your good research won't make a difference. Let's look at how to select the right features and functionality for your product to help your users get things done.

I mentioned in the last chapter that if you used the Behaviour Change Wheel to diagnose ability blockers, you could follow its logic to match those blockers to solution categories. This chapter will step through that process. I'll also show how the different types of solutions can come together in a single experience or product.

Prioritizing Ability Blockers

Chances are, your research to understand your users' ability blockers produced a long list of possible culprits. Depending on how complex the behavior was, it's not unusual to find five, ten, or more ability blockers preventing the target behavior. In all probability, your product won't be able to address more than a few of them. So, figure out which ability blockers to overcome, and which are outside the scope of your work. Then rank order the ability blockers.

Criteria to Consider

The first step in rank-ordering ability blockers is to select criteria to rate each blocker. Go for a blend of idealism and practicality here. The bigger an effect a single blocker has, the more important it is to target it with a design solution. However, you'll also want to consider practical constraints like cost, timeline, resources, and whether going after a particular blocker would take you too far outside your product's boundaries.

Here are some of the criteria you might use to rank the blockers you found. You don't have to use all of them, but pick at least two or three to make sure that you're thinking about multiple angles. Simply going for the blocker you saw most frequently in your data or that was easiest to solve may prevent you from arriving at the most effective design.

- **The prevalence of a blocker.** How many people in your research were affected by the blocker? Something that prevents a large percentage of people from doing the target behavior is much more important to address than something that only affects a few people.

- **The frequency of a blocker.** A blocker that happens all the time will annoy users, even if it's something they're able to overcome. Many users won't have the patience to troubleshoot the same issue continually.

- **The impact of a blocker.** Is the blocker fully *stopping* the behavior from happening? Is it more like a minor inconvenience? The bigger the impact, the more important to address it.

- **The ease of overcoming the blocker.** Not all blockers are easily addressed by a product designer. If it's unlikely that a particular blocker can be removed by your product, then think instead about whether you can do an end run around it (e.g., suggesting cheap alternatives to expensive equipment) or adjust the intended audience for the product (e.g., focusing on people who don't experience that blocker). Other relatively easy-to-overcome ability blockers include revising jargon that people can't understand and making sure that instructions are clear and easy to follow.

- **The cost and effort required to overcome the blocker.** Sometimes a blocker is too difficult to address within the practical constraints of your project. For example, let's say you find that your app is unable to pull data from a specific activity tracker, which stops many potential users from trying it. But to integrate with that tracker would require a complete rewrite of your code base and add two months to your release timeline. It may be better to move forward with that blocker and address it later than to incur the cost and delays of doing it now—unless, of course, the blocker is central to your product in which case, you may just need to bite the bullet and fix it.

- **The alignment with your brand, product, and purpose.** No product can be all things to all people. You may identify blockers that you'd love to remove for your users, but doing so will take you far outside your purview. Those blockers are not a good candidate to address in your product, even though in an ideal world you'd be able to solve them.

Creating a Priority Score

After you've decided on your prioritization criteria, you'll add them to the data grid you built during your research phase so you can score each blocker on your list. You'll give each blocker a score on each criterion. There's no need to use a big scale here—0 to 3 works well, where 0 means "not at all" and 3 is the highest score. Once you've given each blocker a score in each criterion column, add the scores to get a total in each row. See the example grid in Figure 6.1.

ABILITY BLOCKERS + SOLUTIONS NOTES GRID: EXAMPLE

COM-B Category	Subcategory	Evidence for Existence of Blocker	Prevalence (0-3)	Impact (0-3)	Brand Congruence (0-3)	Total Score
Psychological Capability	Knowledge	P2: "I get confused by all the steps I'm supposed to do. What comes first? I always have to look back at the app to check. And then it takes so long to do the workout."	3	3	3	9
	Self-regulation	P5: "I can do the yoga sequence really well if I've got the energy and a quiet space to focus, but as soon as my kids come home it's game over."	2	2	1	5
Social Opportunity	Social support	P14: "I know my husband wants me to be happy, but he has a really hard time when I don't want to sit around with him watching TV anymore. He makes me feel bad about abandoning him and the things we used to enjoy together." P11: "It turns out my good friend also wanted to sign up but was afraid to do it alone. So now we're gym buddies. I know if I skip, I'll be disappointing her, so I go."	2	3	3	8

FIGURE 6.1

Create a grid similar to this one to organize and rank the ability blockers you've identified for your users and target behaviors.

Then sort the blockers on their overall score to rank order them. If you're a quantitative soul or if you have a lot of blockers to sort through, you could create a simple algorithm to add the individual scores directly in your table using Excel or another spreadsheet program.

Select the top few activity blockers to try to solve in your design. There is no "magic number" of blockers to pick; a more complex project might accommodate several, while a small pilot project probably should include only one or two.

In general, accept that you won't be able to fix 100 percent of the ability blockers your users might experience. Focus and simplicity are more likely to produce a high-quality result than trying to do too much. If you're able to identify the most *common* and *significant* activity blockers and design around them, you may find users willing to do some problem-solving of their own for the more minor barriers.

When you've developed your short list of ability blockers to address with your design, it's time to figure out what techniques to use to combat them.

Overcoming Ability Blockers

Most of this chapter will focus on using the Behaviour Change Wheel to choose a category of features or functionality, but it's not mandatory to use this particular tool. If for some reason you prefer not to use the Behaviour Change Wheel, there are other tactics you can use.

You can read behavior change research or case studies related to your target behavior or user population and see what has worked well in other situations. Assess what other digital products in the same subject area have done that was successful (or not) and learn from them. Use your interviews, both informal friends and family conversations and structured research activities, to ask people what's been helpful in the past and what tools and support they wish they had now. If you're talking to people who haven't figured out how to do the target behavior at all, you can always ask how they've made behavior changes in other areas of their life and see if there are any pearls of wisdom there.

Sometimes, you can simply use common sense. For many types of ability blockers, it's not hard to figure how what types of solutions might work. If someone is lacking information, giving them the missing information is a good start. However, "common sense" should be a last resort option, particularly if you don't have a lot of experience designing for behavior change. One of the maddening and wonderful things about psychology is that people will surprise you. Assumptions about how behavior change works aren't always right. Cross-check your guesses about solutions with published research or by talking to people.

For us, we've gone through the hard work of researching our ability blockers using the Behaviour Change Wheel and categorizing them as capability, opportunity, or motivational issues. So, let's use the Behaviour Change Wheel to determine which types of solutions to use for different categories of ability blockers.

The Behaviour Change Wheel offers nine categories of solutions, otherwise known as *intervention functions*, to overcome barriers to performing a target behavior. They are the following:

- **Education:** Providing knowledge
- **Training:** Building skills through instruction and practice
- **Persuasion:** Convincing someone to do or think something
- **Incentives:** Rewards or enticements for doing something
- **Coercion:** Punishments or deprivation for doing the wrong thing or not doing the right thing
- **Restriction:** Making it harder to do an unwanted behavior through rules or other changes
- **Enablement:** Making it easier to do a wanted behavior through rules, tools, or other changes
- **Environmental restructuring:** Changing the physical or social context for a behavior to make it easier or harder to do
- **Modeling:** Demonstrating the desired behaviors to teach by example

Each category of solution is linked to several types of ability blockers, based on research that shows it is particularly effective to remove them. Figure 6.2 shows which types of solutions correspond to which types of blockers.

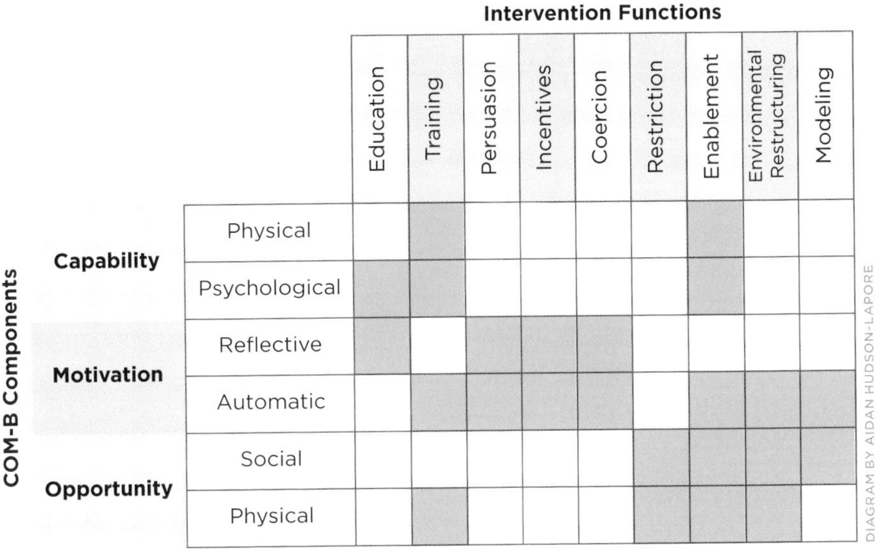

Intervention Functions

COM-B Components		Education	Training	Persuasion	Incentives	Coercion	Restriction	Enablement	Environmental Restructuring	Modeling
Capability	Physical		■					■		
	Psychological	■	■					■		
Motivation	Reflective	■		■	■	■				
	Automatic			■	■	■		■	■	■
Opportunity	Social						■	■	■	■
	Physical		■				■	■	■	

DIAGRAM BY AIDAN HUDSON-LAPORE

FIGURE 6.2

The solid boxes show which types of solutions are most likely to be effective for the corresponding ability blocker category, based on previous research studies. Use this chart to narrow down the types of solutions you might use to help your users achieve target behaviors.

Solving Physical Capability Blockers

If the blocker is physical capability, the solutions tend to fall into two categories: *training* and *enablement*. The concept of training is familiar to most people. It often involves breaking the target behavior into smaller, more manageable pieces that someone can practice until they've developed the ability to perform them all together. A great example of a behavior change program that uses training to overcome physical ability blockers is Couch to 5k.

Couch to 5k is an app that's designed to help people with no running experience get ready to run a 5k distance, or 3.1 miles, in 9 weeks. It has people start slowly—very slowly. Users alternate between short bursts of walking and running. Over time, the running bursts get longer while the walking bursts get shorter, until the person eventually runs 30 consecutive minutes (see Figure 6.3). Creator Josh Clark, from whom you'll hear more in Chapter 8, says one of his primary considerations in developing the program was to remove the physical pain from learning to run. By training people slowly, Couch to 5k helps people build the stamina that removes the physical capability blocker *and* overcomes the aversion to physical discomfort that is an automatic motivation blocker.

Enablement is the other category of solution to physical capability blockers. It refers to tools or strategies that help people overcome a physical limitation. It could take forms like medication to address a health condition that limits ability to do a behavior, a surgical intervention like weight loss surgery, or screen reader software that transforms text to voice for visually impaired people.

Solving Psychological Capability Blockers

If the blocker is psychological capability, such as a lack of knowledge about how to do the target behavior, the solutions might include *education*, advice, or how-to resources to help guide the person through the steps. Here, it's helpful to know specifics about what the person might not know. In general, most people don't have the patience to scan a long document or set of instructions to find what they need, so being able to offer the right tidbit of information at the right time is enormously helpful.

FIGURE 6.3

Couch to 5k overcomes physical activity blockers by breaking runs into short bursts with walking recovery breaks.

TIP KEEP IT SHORT

Financial wellness coach Karen Timmeny tags the educational posts on her blog with how long it takes to read them. She finds that 8–10 minutes is the upper limit of how long people are willing to spend on any one article. If you're including educational content in your product, break it into chunks that users can consume quickly, or they may not look at it at all.

One product that does this very well is TurboTax. On every page, TurboTax has a link next to the data entry field that you can click for more information related to that topic. What's particularly good about TurboTax's method is that as soon as you realize you don't understand something, the right information to answer your question is linked. TurboTax anticipates likely questions and offers the right information where it's needed. And it offers *only* the information that's needed, so people don't have to wade through irrelevant content to find the right bit.

As seen in Figure 6.4, the information also displays right alongside the entry fields so that people don't need to transition between screens (another potential ability blocker if they lose their place, or get distracted by something in another browser tab).

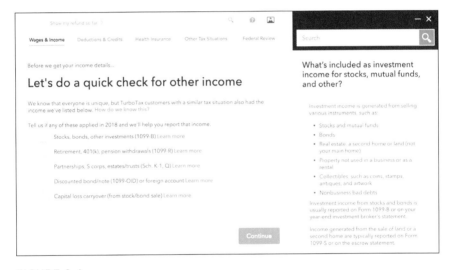

FIGURE 6.4
TurboTax provides users with the relevant education in context, when it's needed. This helps to overcome psychological ability barriers to filing taxes.

For less complicated pieces of information, education might be delivered by footnotes or in-line explanations. For example, if you're building a program that mentions "hypertension," you might include in parentheses that it means "high blood pressure."

Sometimes adding the necessary guidance or education will take you too far outside the purpose of your product. This is where linking to outside resources can be valuable. Just pick reputable ones; for example, with a health behavior change intervention, consider linking to well-known institutions like the Mayo Clinic or the World Health Organization. The information there will be high quality. For an app, there might also be situations where compatibility with other apps can help overcome psychological ability blockers. MyFitnessPal will have an easier time calculating how many calories I can eat after it receives my activity data from my Garmin watch.

Another way to help overcome psychological ability blockers is to use *leveling*, a type of enablement where a task gets broken down into components that start easy and get gradually more difficult. Ideally, the task a user is asked to do presents an optimal level of challenge. It's neither frustratingly hard nor so easy that it's boring. Designing levels means figuring out what is an optimal challenge for a given user at a specific point in time.

Successfully using leveling as a design tactic means making informed choices about how to chunk material and organize it so that most people would agree that the earlier levels are easier. With some types of behavior change areas, like learning a language, the chunking isn't too mysterious. Other times, it may come down to the preferences and skills of your users.

Consider someone who's just received a new diagnosis of type 2 diabetes. The medical advice this person's likely gotten includes adjusting their diet, being regularly physically active, taking any prescribed medications exactly according to instructions, testing blood sugar several times a day, possibly adjusting medications based on the blood sugar results, and oh, visiting a whole slew of medical specialists to prevent complications. None of these behaviors are simple on their own, but people are asked to take them all on at the same time. It's a recipe for frustration.

You can use a type of leveling strategy to help people with these kinds of significant, multipart behavior changes. Have them pick just one behavior to focus on first. Unless there's a crucial medical reason why a particular behavior should be prioritized, it's better to let the person have a say about which one to tackle. That makes the change more likely to stick. Plus, most people will choose a behavior they feel more confident they can handle, and if they can experience some early successes, that builds an appetite to try another, less comfortable change later.

So for this newly diagnosed person, a behavior change intervention might focus first on taking medications, and then as the person gets better at it, start slowly adding in other new behaviors. New behaviors are usually hardest at first, and gradually become easier as people practice and troubleshoot. "Level up" by adding new behaviors to try after the first ones have become more routine.

Overcoming Social Opportunity Blockers

When the ability blockers are related to social opportunity, the tactics to address it fall into the categories of *restriction, enablement, environmental restructuring,* and *modeling.* These tactics focus on redesigning the environment to limit exposure to people who make the behavior change harder, and increase exposure to people who facilitate it.

Let's look at the case of someone who wants to quit smoking. It's not unusual for smokers to be friends with other smokers. And quitting is a lot harder when you see other people lighting up, smell the cigarette smoke on them, or hang out in the situations where normally you'd smoke yourself.

So the tactics the soon-to-be-former smoker might try to overcome these social opportunity blockers center around changing the situation, which is what the VetChange smoking cessation app in Figure 6.5 recommends. Changing the situation could include hanging out with friends who don't smoke or going to smoke-free locations, especially in the first weeks of the quit attempt when cravings are at their strongest (restriction/environmental restructuring). They could plan gatherings at places where smoking wasn't allowed or ask to be seated in the nonsmoking section (environmental restructuring). They could consider wearing a nicotine replacement patch or chewing nicotine gum to help resist cravings, and ask the smokers in the group to support the quit attempt by not offering cigarettes (enablement). And they could

FIGURE 6.5

VetChange recommends that users experiencing an urge to smoke look for ways to change their situation, which often includes removing social opportunity blockers.

invite a nonsmoking friend (or better yet, a friend who had success-fully quit smoking themselves) to hang out more often, and pay attention to how they handled the situations that usually triggered smoking (modeling). These tactics all reshape the social environment in a way that makes it easier to stop smoking.

Overcoming Physical Opportunity Blockers

Similarly, the tactics to reduce the effects of ability blockers related to physical opportunity focus on reshaping the environment in one way or another (except for training, which helps people negotiate the environment more successfully). They fall into the categories of *restriction, enablement,* and *environmental restructuring.* For eating less and more nutritious food, this might look like the following: not keeping tempting junk food in the house (restriction); having pre-cut fruit at eye level in the fridge in a clear container so that it's very visible (enablement/environmental restructuring); and writing out a weekly meal plan to guide shopping and ensure that the right ingredients are available (enablement/environmental restructuring).

One example from the physical world comes from an increasingly popular setup in public parks (see Figure 6.6). There has long been signage encouraging people to clean up after their dogs. One reason people might not do that is if they don't have anything to pick up the poop with. So, the sign incorporates a small container of free poop bags. Ability blocker, vanquished!

"Not enough time" is another barrier that might fall under physical opportunity blockers. Breaking an activity into shorter chunks, like the Johnson & Johnson Official 7 Minute Workout® app in Figure 6.7, is a useful strategy to overcome that blocker. As a bonus, this app also lets users customize the intensity and exclude hated exercises, so it also tackles blockers from other categories.

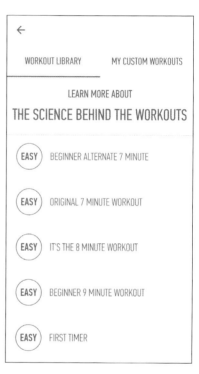

FIGURE 6.6
Putting free poop bags and a trash can alongside a reminder sign ensures that pet owners won't experience common physical opportunity blockers associated with dog walking.

FIGURE 6.7
The Johnson & Johnson Official 7 Minute Workout® app offers a high-intensity interval workout that can be done in seven-minute segments to overcome the barrier of not having enough time to exercise. Ideally, a user will repeat the seven-minute workout several times in one session, but the app permits them to do as much as they have time for.

Overcoming Reflective Motivation Blockers

The last sets of ability blockers are related to motivation. First, there are ability blockers related to reflective motivation, the more planful type of motivation associated with people's goals, values, and desires. The categories of solution for overcoming reflective motivation blockers include *education, persuasion, incentives,* and *coercion.*

Education and persuasion are both ways to work on changing someone's mind so they reorganize their priorities to better align with what you'd have them do. This is often less about having people change their goals and more about helping them see how a behavior change can support the goals they already have, similar to how you can help guide people through choice by aligning options with their values. So, for example, your product might show people a calculation of how contributing money each month to an investment account will allow them to reach their goal of an amazing beach vacation in two years. That's education because it imparts new information, and it also counts as persuasion because you're linking the warm feelings associated with dreaming of the vacation with the decidedly less emotional idea of saving money.

Education can also be direct. For example, in our construction site safety research, we saw that most workers had been taught what good safety behavior looked like. Critically, most of them had also been taught that *construction professionals practice good safety behavior.* When a pipefitter's mental model of what a successful pipefitter does includes following safety rules, then following those rules has a higher priority.

Incentives and coercion refer to rewards and punishments. I've worked on products where my clients have associated financial rewards or punishment with completing program milestones. For example, many digital health programs are implemented by health insurance providers so members who complete them receive a rebate (incentives) or perhaps don't see their premium costs rise (coercion).

As a rule of thumb, try not to attach financial consequences to completing behavior change activities. Paying people to do something reduces their natural enjoyment of the activity over time. There's fMRI data that shows the brain's pleasure centers don't fire as strongly when someone is playing a video game if they're being paid to do it! The good news is that these interest-zapping effects don't seem to be permanent, but many behavior changes need to "stick" in the first few months to become lifelong patterns. If someone is being rewarded for a behavior while it's still new and then that reward stops, there's a high risk that the behavior stops, too.

However, there are other types of rewards and punishments that can be valuable tools for behavior change. One type of self-administered reward is called *temptation bundling*—for example, people who are trying to do a disliked new behavior (working out) pair it with something they enjoy but don't get to do often (watch *Real Housewives*). Knowing that the TV on the treadmill is the only one where they can indulge the guilty pleasure can keep people interested in visiting the gym.

Similarly, people can decide to reward themselves after they've done a new behavior for a while (buying new shoes if I go six weeks without a cigarette) or punish themselves for not sticking to a behavior change plan (I have to put $5 in a swear jar every time I drop an F bomb). Linking these rewards to the target behavior whenever possible creates a positive cycle. A new pair of running pants is a great reward for sticking to a workout schedule if it gets the person excited to wear them to class.

Self-administered punishments exist, too. Pavlok is a wristband that users program to deliver an electric shock when they engage in a bad habit like smoking or eating chips. The inventors claim that the shock effectively deters people from the behavior over time, eventually breaking the habit. Users can control what triggers the shock and adjust its intensity within a range (see Figure 6.8). It remains to be seen whether self-administered electric shocks will catch on as a behavior change technique.

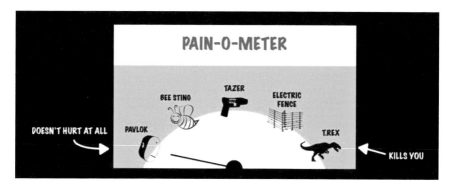

FIGURE 6.8
Somehow the Pavlok being less painful than a T. Rex mauling isn't all that reassuring.

Overcoming Automatic Motivation Blockers

The solution categories for overcoming blockers related to automatic motivation are *training, persuasion, incentives, coercion, enablement, environmental restructuring,* and *modeling.* Remember that automatic motivation is less explicit than reflective motivation. It can come from the cultural context in which people live, the emotional context in which behaviors happen, and the habits people have developed over time through repetition.

The way to combat ability blockers related to automatic motivation is to chip away at them by offering alternative ways of thinking about the world. With our construction safety research, we encountered a number of workers who either believed that nothing they could possibly do would prevent a freak accident if it was going to happen, *or* that the odds of something happening to them were so low that they didn't need to worry about it. Both of those belief patterns were automatic motivation ability blockers that led to fewer safety behaviors. What seemed to overcome it was when workers heard firsthand accounts of a serious injury or fatality, or worse, witnessed one themselves. That experience helped jolt them out of an automatic motivation pattern that wasn't serving them.

Of course, you don't want to expose people to traumatic events in order to get them to change their behavior. But you can offer a gentler version of those sorts of transformative experiences by sharing stories and examples from people who have successfully made a behavior change (modeling, persuasion). As people encounter more evidence that differs from their starting belief set, their cultural expectations shift, and the new behavior begins to seem possible.

The counterexamples to what a person expects can also serve as incentives and coercion. Recall that automatic motivation is influenced by the costs and rewards associated with doing a behavior, like the way others might react. If examples show that people who do the target behavior are positively regarded or treated well, that can change the reward calculation about whether the behavior is worth doing.

Training, aka *practice*, is another way to reshape people's expectations gradually. As a person tries new behaviors and sees the actual consequences (which may not be what they expected), their belief structures about the behavior may begin to change. For example, when I was with Johnson & Johnson, our Human Performance Institute used training to get CEOs to take short walking breaks during the workday. Many of them started off thinking that walks would be a waste of their valuable time and add more stress to a busy schedule.

The trick was getting the CEOs to monitor their energy levels. What these behavioral experiments showed was that short bursts of movement made the CEOs feel more focused and energetic afterward. This let them get more and better work done. They spent five minutes, but gained more productivity in return. Suddenly, walking didn't seem like a waste of time but like a "productivity hack."[1] When people practice a low-stakes behavior like a five-minute walk and pay attention to the outcomes, they gradually change their belief system around the behavior.

Enablement and environmental restructuring can target automatic motivation blockers by interfering with the habitual behavior patterns people have developed. In psychology, the word "habit" refers to something that is done without conscious thought in response to a cue in the environment. The big key is "without conscious thought." In daily life, we call a lot of things habits that aren't really, but for behavior change work, it's important to know the difference.

To drive user behavior change, think about tools and environmental changes that break unconscious patterns. You might add something to their environment: reminders or "habit pairing" are great examples of this. If someone usually hits the couch instead of the gym after work, setting an alarm on their phone for 6 p.m. could disrupt the pattern. Post-it Notes left in key habit locations are an inexpensive tool to get people to pay attention to what they're doing.

Or you could remove something from the environment that's prompting the behavior. If a compulsive potato chip–eater wants to stop eating so many fried snacks, a helpful solution is to make sure the chips aren't in the house. Once a person has to make multiple effortful decisions to do something (put on coat, leave house, go to store, locate and buy chips, come home, eat chips), it's no longer possible for it to be the result of automatic motivation.

---

1 I wanted to use the eye-rolling term "productivity hack" at least once to increase the appeal of this book for tech startup bros. Hey guys!

Translating Intervention Functions to Solution Features

It's time to update your COM-B grid. (See Figure 6.9 to translate the intervention functions for your top blockers into solution features.)

ABILITY BLOCKERS + SOLUTIONS NOTES GRID: EXAMPLE

COM-B Category	Subcategory	Evidence for Existence of Blocker	Prevalence (0-3)	Impact (0-3)	Brand Congruence (0-3)	Total Score
Psychological Capability	Knowledge	P2: "I get confused by all the steps I'm supposed to do. What comes first? I always have to look back at the app to check. And then it takes so long to do the workout."	3	3	3	9
	Self-regulation	P5: "I can do the yoga sequence really well if I've got the energy and a quiet space to focus, but as soon as my kids come home it's game over."	2	2	1	5
Social Opportunity	Social support	P14: "I know my husband wants me to be happy, but he has a really hard time when I don't want to sit around with him watching TV anymore. He makes me feel bad about abandoning him and the things we used to enjoy together." P11: "It turns out my good friend also wanted to sign up but was afraid to do it alone. So now we're gym buddies. I know if I skip, I'll be disappointing her, so I go."	2	3	3	8

What will the education, training, enablement, etc. look like in your particular product?. In the last column, write out some thoughts about what solution features would use the intervention function listed to solve the ability blocker in that row. You'll want to make sure that your ideas work within the boundaries of your product.

Intervention Function	Solution Feature
Training	Step-by-step videos breaking the workout down into bite-sized chunks
	Offer a list of specific ideas of when/where/how people can find quiet space for yoga in a busy home. Interview moms and use their stories.
	Coach people in asking for support; generate list of benefits to new behavior that can help people who miss the old way of life see some positives?

FIGURE 6.9
Use your notes grid to document ideas for how to bring each solution category to life in your product.

Don't suggest lengthy physiology articles if you're building an app to make exercise simple and fun for people. This is also where you may make some first slices based on the practicality of a solution feature. Video may not be right for you if no one on your team has video-editing capabilities and your release date is looming.

It's up to you whether you want to add ideas for the lowest-ranking ability blockers. I usually do; sometimes there's a quick and easy solution to a low-priority blocker that you might as well implement. It can also provide reassurance that the solutions you *are* implementing are the more important ones and begin to fill your release backlog with good ideas.

At this point, you may have only a high-level idea of what the solution features will look like. That's okay. In fact, in the sample grid, you'll see some of the ideas are really questions about whether or not something could work. The next step will be for the team to talk through that. Detailed requirements come later.

Tried and True Ability Enhancers

These extended examples describe some common ability blockers and solutions. With these examples, it should become clear that behavior change design usually involves blending multiple types of approaches to enhancing people's ability. That's because it's very rare that there are only one or two straightforward ability blockers at play. You'll likely have realized that if you've thought about applying the COM-B system to a problem area in your own work.

You also may have picked up on some of the fuzzy lines between the types of blockers. Is being judged negatively by your friends for a behavior a social opportunity problem, or is it an automatic motivation one? Well, it could be either or both, depending on how the person experiences it. However, because the Behaviour Change Wheel system is based on so much research, usually these sorts of overlapping blockers have shared solution categories. It matters less that you get the categorization exactly right and more that you choose a solution that works.

From Opt-In to Opt-Out

Opting into something often takes effort. You have to indicate interest. You may have to fill out forms or make decisions about what your participation will look like. Sometimes, it's not clear what opting in even means. Will this be a one-and-done commitment, or are you going to ask for my credit card on the next page?

As a rule of thumb, *action is harder than no action.* Since people often do what's easiest, how can you design it so that inaction becomes a good thing?

A technique to increase enrollments in some types of programs is flipping from opt-in to opt-out (also called *automatic enrollment*). People are automatically signed up to be a part of the program unless they make the effort to say no. This has been used by employers to get employees to sign up for 401(k) retirement savings accounts. An opt-in 401(k) tends to have fewer people signed up than an opt-out plan. According to Vanguard, among their customers, the number of opt-out plans offered to employees has grown by 300% in 10 years.

The opt-out technique has also been applied to getting people to register as organ donors. The model still in use in most places in the U.S. is that people have to deliberately register as an organ donor. (I did it as part of getting my driver's license.) Research shows that in countries where people have to opt-in to organ donation, around 15% of people do it. Some European countries have moved to an opt-out model, and they're seeing much higher numbers. For example, in Austria the rates of people signing up to be organ donors are over 90%. Interestingly, people in opt-in countries think organ donation is a much bigger deal than people in opt-out countries. Something about "everybody doing it" seems to minimize the emotional impact of the topic.

An example where opt-out is underused to drive behavior change is hotel sustainability programs, usually involving reusing towels and sheets or giving up housekeeping during your stay. Hotels usually require guests to opt in to these programs by hanging a card on their doorknob or leaving it on their pillows. In one of the Swissôtel examples in Figure 6.10, they actually ask you to call the front desk![2] I'm someone who's very motivated to conserve natural resources, and yet I rarely remember to leave the card in the right place. I'm certainly not likely to make a phone call. Switching to an opt-out could be very effective for hotel chains who truly want to reduce water and energy usage.

---

2 I am very grateful to millennials for killing the phone call.

Like many hotel chains, Swissôtel allows guests to opt in for more sustainable housekeeping practices, but requires them to do a little work to make it happen.

A word of caution: Opt-out gets people started in a program, but won't be sufficient if you need them to keep taking action. In the example of the 401(k)s, although many more people end up enrolled with an opt-out plan, they tend to save less money than the people who opted in. They're not paying the same type of attention and making decisions to increase their contributions. Instead, they just save the minimum amount that's part of the opt-out. You could argue that it's better than saving nothing at all, but it also leaves a lot of room to help people optimize their savings.

Personalizing for Ability

Many behavior change products focus on getting people to do something differently from now on, not just a one-time action. For all of those projects, the design needs to take people from wherever they are now and bring them closer to an ideal state. Often, there's a lot of ground to cover between the two. Personalizing the experience based on their current state is an excellent way to coach people without running into as many ability blockers.

Duolingo is a free language-learning program that also happens to offer a masterclass in designing for behavior change. Its personalized approach allows it to handle a number of ability blockers elegantly (see Figure 6.11). The content is chunked into short lessons of 10–20 questions. The lessons are organized from the most basic, simple greetings and short words, to most complex, which for Spanish is conjugating verbs in the conditional perfect tense.

Not only do the lessons become increasingly hard as you go, they build on each other. Lesson 3 pulls in vocabulary from Lesson 2. And if you go back and practice an earlier lesson after progressing, the earlier lesson will include some of your new words. Plus, users with existing language skills have opportunities to opt into more advanced lessons, which keeps them from

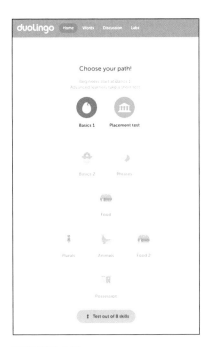

FIGURE 6.11

Duolingo displays the available language lessons in an orderly sequence so that they build on one another over time with easier lessons first. Advanced learners can opt into more difficult lessons.

getting bored (a recipe for disengagement). Duolingo offers one of the best examples of *leveling* in a digital product to help overcome *psychological capability blockers*.

Chunking the lessons into such bite-sized pieces helps with another potential ability blocker—time. Someone can whiz through a Duolingo unit in under five minutes.

Another tactic Duolingo uses to overcome ability blockers is formatting the individual test items to enhance learning and recall. Questions include an audio component with the correct pronunciation of the word or phrase being learned. If learners need a prompt, mousing over the word will show its translation in a pop-up or trigger a voice-over. Some early lessons also pair images with the questions to help build an association between a word and its meaning (see Figure 6.12).

Users also have the option to enable their microphone, in which case some questions will test their speech. Ideally, a new language learner will find opportunities to speak, but a Duolingo user might be using the program in situations where speaking aloud isn't realistic. I sometimes do a Duolingo lesson on the subway, for example, and I've done the occasional lesson in my open office area. People would definitely notice me trying to enunciate in another language. That's an automatic motivation blocker for me.

Which of these is "map"?

ちず   かぜ

よる   ぎんこう

CHECK

FIGURE 6.12
Pairing new vocabulary words with images helps learners recall them more easily and overcomes some of the ability barriers associated with language learning.

Watch Out! Ethical Considerations in Designing for Ability

When identifying and solving for ability blockers, it's important to come from a place of empathy as much as possible. Designers are often experts in the areas where they try to get people to change their behaviors. That means they may not remember what it's like to be new at this type of thing. They may be naturally gifted at something that's hard for others. Or they may have experienced privileges that helped them along the way, but that their users don't enjoy.

From a position of empathy, the response to an ability blocker is to find a solution. If you can't adopt that empathy, then you might do one of the following:

- **Ignore the blocker:** "That shouldn't be a problem, so I'm not going to do anything about it."

- **Minimize the blocker:** "You may feel like this is a problem, but it shouldn't be. Just get over it."

- **Talk down to the user:** "That workout seemed really hard for you. You're overreaching." (The word "overreaching" is actual feedback from my Garmin!)

None of these approaches will engage people. Engaged users are ones who feel understood, not blown off by design teams.

It's also important to give users the option *not* to be engaged. Although behavior change designers look at motivation from multiple angles, at some level it boils down to "Does the person want to do this thing?" The answer very well may be no. Even though behavior science offers some tools to get people to do things they don't want to do, the point of behavior change design is usually to get someone to do something differently in a long-lasting way. People don't maintain new behaviors they don't really want to do. Forcing change backfires in the long run.

Out of all of the ability enhancers in this chapter, the two types with the most potential to be misused with respect to forcing people to do something are incentives and coercion. I've tried to give examples of incentives and coercion that respect your user's autonomy and right to say no.

Sometimes, people feel that if they're offering a reward, it's an unqualified good thing. But counterintuitively, rewards may reduce people's free choice. If you offer people too much money to do something, it becomes unreasonable to say no. They have no real choice. Case in point, many workplace wellness programs like the one you read about in Chapter 1 offer financial incentives that make health insurance more affordable. As a result, people sign up for the programs and do what's required for the reward, but nothing more. As much as possible, include only token rewards that support the target behaviors in your incentives plan to sidestep the ethical and practical issues of a rewards-heavy approach.

The Upshot: Pick the Right Solution

Designing a behavior change intervention that works is all about selecting the features that most effectively combat any blockers your users face while amplifying any boosters they encounter. Using a model like the Behaviour Change Wheel to trace a line between the ability blockers you've identified and the intervention features you're building will help you maximize your odds of success.

Choosing a solution type means making other hard choices, too. No one intervention will solve all of the ability blockers a user might encounter, so it's important to prioritize and make choices that are aligned with what you're able to offer people. Fortunately, there are solution types that address multiple types of ability blockers, so using those when you can will help you reach the greatest number of users.

Selecting the right product features to help people achieve their goals requires a thoughtful interpretation of research insights and strategy to design—exactly the nexus of activities where Sheryl Cababa works. If you've heard Sheryl speak, you know one of her missions is to help designers ask the right questions, which often may be less "How should we?" and more "Should we?" Behavior change design offers an opportunity to do that sort of critical reframing at the point where research insights are translated to feature requirements. Here are some of Sheryl's thoughts on navigating that challenge.

How can you select effective intervention tactics?

You can't go out of the gate like, "Okay, we're going to do an app that's going to do this thing," because that might not be necessarily the best intervention. It's better to start with the outcomes that you want and then work backward. There's a philosophical alignment that needs to happen at first with whomever the key stakeholders are on a project. If there's some flexibility around the shape that the solution might take, that gives you a lot of leverage to actually design for outcomes rather than designing for a technology.

Oftentimes, the needs of the organization tend to outweigh the needs of the end users or the end beneficiaries of our products. Constantly keeping that lens on audience and who's going to benefit or potentially be harmed by your product is probably the most important thing to be cognizant of throughout the design process. It might be that you have individual users of your product, but you also have an ecosystem of people who might experience secondary harm. Having an understanding of that ecosystem helps specifically avoid unintended consequences.

How do you know if your product might have an unintended harmful effect?

One of the things that we have done at Artefact is to develop the Tarot Cards of Tech,[3] a tool for designers and technologists to use to consider potential unintended consequences of their work. They're a series of prompts meant to be used within the product design and strategy process to help designers consider what happens at scale. What would using your product too much look like? What would happen if 100 million people were using your product? As we think about questions of equity, inclusion, and access, who is your product leaving out, and how do you better design for them? How would bad actors use your system or digital product to potentially harm other people?

---

3   Available to download as a PDF at http://tarotcardsoftech.artefactgroup.com/

I just don't think there's enough diversity still in the industry to be able to look through different lenses of people's diverse experiences, especially on platforms and digital products. When you have people thinking about these things in abstract ways rather than experiencing it in their own skin, then it's sometimes hard to grasp the weight of those potential unintended side effects.

What is the value of a systems lens?

Product teams are really siloed. It's hard for them to see the big picture of what they're doing. It's easy to get into this myopic state of just working on the little thing you're working on, then forgetting that there's this big picture of abuse on your platform. You're working on features where you don't actually know what they're ultimately going to contribute to from a strategic or business model lens. Designers being able to use the systems lens collectively about the product at large is really important. You should really understand the business model of whatever it is you're working on so you know how your company makes money and understand your role within that.

What can designers do if they're not comfortable with a solution?

Know your values and stick with them. Try to find ways of working them into the products and services that you're designing for. The opposite side of that is be willing to interrogate your complicity. If you're asked to design features, ask yourself about whether you would want to be at the receiving end of these kinds of behavioral manipulations. Not only that, but call them what they are. If you, for example, are working on something that increases data collection, would you think about it differently if you called it "surveillance"?

It's too easy an answer to say, "You should just leave." That's a position of privilege. You have to have incremental and actionable tools and methods to use to try and do the right thing within the scope of what you're doing. You can help people take a broader lens without necessarily telling them to light everything on fire.

 **Sheryl Cababa** *is an Executive Creative Director at Artefact in Seattle. Prior to that, she worked with frog and Adaptive Path. Sheryl's clients span industries from health care, education, and technology and include many of the world's leading companies.*

CHAPTER 7

Harder, Better, Faster, Stronger

Designing for Growth

You've identified a target behavior and uncovered what's stopping people from doing it. You've selected solutions to overcome those blockers. Now you've got people using your product to change their behaviors. How do you help them build and maintain momentum over time?

Because people thrive on progress, the path to a new behavior can be thorny. It's littered with failed attempts and setbacks, as anyone who's tried to make a significant change in their own behavior knows. Changing people's behavior in a meaningful way requires taking the sting out of failures and helping people feel like they're moving toward success.

That's why effective behavior change design includes breaking hard goals down into more achievable pieces and making sure those pieces fit the person's sweet spot in terms of difficulty and interest. Designers can add interest and emotion to the behavior change process to keep people invested in a long journey.

Remember that one of the key psychological needs people have is to feel a sense of competence, or learning and progress. People are wired to be extremely sensitive to indications of whether they're succeeding or not. Whether it's noticing the micro-expressions on someone's face during a conversation and adjusting their tone of voice to avoid angering them, seeking out points and grades on school assignments so they can make sense of how they did, or comparing their race times to other people in their age group to see if they're actually any good at running, people are always picking up on and making use of feedback in their environment to understand themselves. Good feedback helps users stay the course.

Motivating Growth

Behavior change is about progress over perfection. No, you didn't suddenly fall into a cheesy Instagram motivational quote rabbit hole. There's actual science behind this. Research shows that when people have *growth mindsets*, which include a focus on improvements and a belief that effort will translate into results, they're well-equipped for behavior change. On the other hand, people with *fixed mindsets* focus on achievements and believe that results come from natural talent. You can imagine how someone who's okay working hard on small steps toward a big goal might do better with behavior change than someone who thinks that if it doesn't come easy, it may not be worth doing at all.

Fortunately, people aren't stuck with their default mindset. The way you design the goals of your behavior change program can nudge people toward more of a growth mindset. At the same time, you can structure the behavior change activities so that they are truly achievable for people, given the right supports.

Frame Challenges Positively

People with a fixed mindset perceive difficult tasks as a sign that they're not cut out for that activity. Can you provide context and framing that helps battle that way of thinking? For example:

> This is a really hard thing to do, so don't worry if you can't do it at first.
>
> Most people need to try this many times before they can do it.
>
> This is a really challenging goal, so we're going to tackle it in small pieces.

If people do fail, offer encouragement. Suggest they try again later, offer ideas for approaching the task differently, and remind them that change takes time. You may also be able to reframe success; perhaps users didn't achieve one goal, but did make progress on a different one. Point it out.

If people succeed, congratulate them! But be careful to give praise that supports a growth mindset and not a fixed one. Saying something like "You're a beast!" implies that success was due to something about the person—exactly what a fixed mindset believes. Consider instead reinforcement like "Your hard work paid off!" or even a simple "You did it!"

It's All About the Milestones

People love seeing evidence of their own progress. When they do notch a win, it increases their *self-efficacy*. Self-efficacy is confidence that they can do something, and people who have more of it are more likely to try new activities until they succeed. And then those new successes continue to feed self-efficacy. Figure 7.1 shows how Habitica, a behavior change program to build positive new habits, gives its users an instant self-efficacy boost.

Good behavior change design offers people lots of opportunities to experience success and build self-efficacy. Breaking the major behavior change goals of your program down into smaller, more easily achieved milestones gives people many chances to succeed along the way. It also helps them to be more resilient if they don't get something right the first time.

If the milestones toward a goal build logically on one another, then you're using the leveling technique discussed in Chapter 6. They can be totally linear, such as adding more duration or intensity to the same activity over time. Or milestones can offer people options along the way to follow different branches, so that different types of activities lead to the same behavior change goal.

Milestones can also send people on "side quests"—tasks that support the overall behavior change program but may not

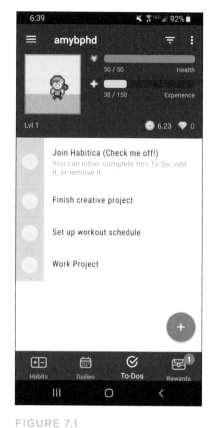

FIGURE 7.1
Habitica's users can check "Join Habitica" off their to-do list right away. This combines necessary onboarding with a self-efficacy boost.

directly relate to the main goal at hand. This can be a good way to handle items that are important to get done but don't need to come at a specific time in the behavior change journey. For example, people beginning marathon training will want to register for a race, but as long as they have a rough target date, they can do that before or after setting up a training plan and going on the first few runs.

Figures 7.2 a, b, and c show examples of programs that use different types of milestone structures to get users to a behavior change goal.

Summary | Program Details | Training Schedule | Additional Programs | Reviews

Week	Mon	Tue	Wed	Thu	Fri	Sat	Sun
1	Rest	3 mi run	2 mi run or cross	3 mi run	Rest	30 min cross	4 mi run
2	Rest	3 mi run	2 mi run or cross	3 mi run	Rest	30 min cross	4 mi run
3	Rest	3.5 mi run	2 mi run or cross	3.5 mi run	Rest	40 min cross	5 mi run
4	Rest	3.5 mi run	2 mi run or cross	3.5 mi run	Rest	40 min cross	5 mi run
5	Rest	4 mi run	2 mi run or cross	4 mi run	Rest	40 min cross	6 mi run
6	Rest	4 mi run	2 mi run or cross	4 mi run	Rest or easy run	Rest	**5-K Race**
7	Rest	4.5 mi run	3 mi run or cross	4.5 mi run	Rest	50 min cross	7 mi run
8	Rest	4.5 mi run	3 mi run or cross	4.5 mi run	Rest	50 min cross	8 mi run
9	Rest	5 mi run	3 mi run or cross	5 mi run	Rest or easy run	Rest	**10-K Race**
10	Rest	5 mi run	3 mi run or cross	5 mi run	Rest	60 min cross	9 mi run
11	Rest	5 mi run	3 mi run or cross	5 mi run	Rest	60 min cross	10 mi run
12	Rest	4 mi run	3 mi run or cross	2 mi run	Rest	Rest	**Half Marathon**

a

b

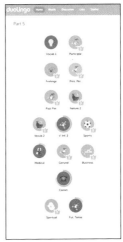

c

FIGURE 7.2

Hal Higdon's novice half marathon training program (Figure 7.2a) breaks the goal of running a half marathon into linear milestones: weekly long runs that slowly increase in distance. Janssen's CarePath diabetes management program (Figure 7.2b) follows a "crooked path" of milestones where users skip from topic to topic each week, but every milestone gets them closer to the ultimate goal of managing their diabetes well. And Duolingo (Figure 7.2c) offers side quests and branching paths, so users have choices about which language topics to delve into but still arrive at being able to pass the unit tests.

Generating meaningful milestones for a goal behavior can be challenging. Here are three ways to do it.

Read published research. Studies on people who have successfully changed a behavior will give you ideas about milestone goals for your users. For example, the National Weight Control Registry follows thousands of people who have lost weight and kept it off. One thing a lot of them have in common is that they weigh themselves every day. So a daily weigh-in is a reasonable milestone for someone losing weight.

Talk to subject matter experts. Involve people who know a lot about your target behaviors and ask them for input on good milestone goals. These may be paid consultants for your project, but if you don't have that kind of budget, try sending an email or even a tweet. Ask a focused question like "What are two things my users should be able to do by the time they achieve this goal?"

Ask your users (or people like them). User research will help you learn what kind of milestones people typically hit on their journey toward a goal like yours. Ask "What kinds of things helped you feel like you were making progress?" and "When did you feel like you'd leveled up?" After a while, you should start to see patterns in people's answers. Those patterns will shape your milestones.

Flow and Grow

Flow refers to a state of engagement where a person is so absorbed in what they're doing that time passes by unnoticed. People experiencing flow are completely focused on their task. Flow feels good. Neurological research shows that people's prefrontal cortexes quiet down during flow, which means their inhibitions are lowered and their brains may be able to work more creatively.

You may have experienced flow while reading a great book, playing a favorite video game, or putting the finishing touches on a giant jigsaw puzzle. Someone in a state of flow is truly engaged with what they're doing. You want to give your users as many moments of flow as you can.

Here's how.

Flow happens when people experience optimal levels of challenge. That's when a task is in the sweet spot in between being too easy (boring!) and too hard (frustrating!). Within that between space, there's room to offer more challenging tasks that push the curve upward. These are behaviors that are a stretch for someone, but they have the skills and ability to do with effort.

Alternate these challenges with easier tasks for people to rest and consolidate their learning. Just like with building muscles, people build their capabilities by alternating work and recovery. The recovery is essential for progress.

Over time, the flow space will bend upward as people become better at their new behaviors. This illustration in Figure 7.3 from Jane McGonigal's game *SuperBetter* shows how repeated quests help players increase their skills over time.

FIGURE 7.3

As *SuperBetter* players complete increasingly difficult quests, their skill levels increase so that they can tackle harder and harder challenges.

Don't Be Boring

Boredom isn't only bad because it keeps people from experiencing flow. It can also lead them to disengage with the behavior change experience entirely. You don't want your products to be boring!

Being too easy isn't the only way to be boring, either. Being repetitive is also boring.

Avoid asking people to do exactly the same thing every day for many days in a row. Yes, some behavior change does require repetition.

Developing a habit (in both the true psychological sense and the colloquial one) means doing the same thing repeatedly until you don't even think about it anymore. But there are ways to add some spice:

- **Mix up the milestones.** It's not always just about intensifying the same goal. You can also focus on other aspects of the behavior. For example, a runner doesn't need to keep going longer and longer distances indefinitely. Instead, they may focus on speed or different styles of running (trail, road, track, relay), or a goal like running a race in each state. The screenshot from Strava in Figure 7.4 shows the many different types of fitness challenges a user can try.

- **Vary the content around the behavior.** If you're asking someone to weigh themselves each morning, can you phrase the request differently day to day? Or pair it with a piece of interesting information?

- **Celebrate streaks.** If repetition is part of the magic, then reward it! Congratulate users who complete actions without skipping any (or many) days. Seeing an unbroken streak of successes can be very rewarding.

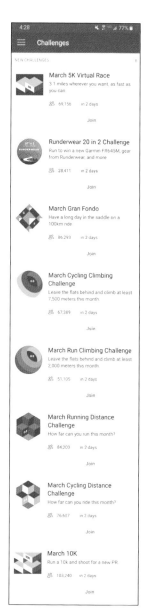

FIGURE 7.4

Strava allows users to select from a wide variety of fitness goals and challenges. This variety helps appeal to more users and can prevent people from getting bored over time.

- **Be brief and be gone.** Once users are on their way with behavior change, they probably don't need a ton of explanation and context. Let them perform or log their behaviors with a minimum of fuss so they can get on with life.

- **Don't be afraid to have a sense of humor.** If you absolutely must ask people to do the same thing repeatedly, own up to it.

(Re)Connect with Purpose

People are more likely to pursue goals that they choose themselves and that align with their core personal values. In the thick of working toward a behavior change goal, people might lose sight of why they started in the first place. The more you can remind them why they care about this behavior change, the better fortified people will be to continue moving through milestones. Some ways to do this include:

- Reiterate what users have already told you, if they've shared information about why this goal matters to them

- Reference universal shared values that are likely to resonate with your users; "Achieving your goal will help you be more effective in your work because . . ."

- Use rhetorical questions in your content to get users thinking about the meaning behind their goals. "Why did you want to try this in the first place? What will achieving this goal mean for you?"

- Let your users add reminders of why their goal is important; a classic tactic for people trying to quit smoking is to keep a photo of a loved one where their cigarettes used to be. Digital makes this kind of visual reminder easy

The longer it takes to reach a goal and the harder it is to get there, the more important it will be to help your users reconnect with their purpose along the way.

The Feedback Playbook

A key tool to move people through the milestones toward the behavior change goal is feedback. That is, you need to create motivating feedback.

You *could* let people construct their own interpretations of how they're progressing toward a goal. But this is asking for problems. People may not have the knowledge to gauge their own success. They may be so excited to do well that they overlook poor performance. They may be perfectionists who don't see small successes as worth noting. Don't get into these issues. Instead, your design should deliberately and thoughtfully build feedback that helps guide people where you want them to go.

This means knowing what makes feedback effective and what to avoid. Good feedback is like a good goal. Recall that SMART goals are specific, measurable, achievable, results-oriented, and timely. Similarly, good feedback is specific, has some objective component, gives meaningful direction for next steps, guides people toward a goal, and is delivered near enough in time to the action so that people can learn from it and adjust their behavior accordingly.

Good feedback is also delivered at multiple levels, especially for complex behavior. This means that feedback should consider both people's most recent behavior and their progress over time. Competence is about feeling like little actions add up to something bigger. Feedback can help people get that impression that all of their individual actions are part of a larger whole.

Good feedback doesn't just focus on what needs improvement. Usually, people already know when they're doing badly. They may even avoid seeking feedback because it feels lousy to hear they have messed up. Negative feedback can be helpful because it gives information about what not to do and how to do better, but it's usually not feel-good news. Mixing in a dose of positive feedback can help make the overall experience less painful for people. Plus, being caught doing something right gives a sense of growth and can boost confidence to try something new.

The final step of good feedback is offering a new call to action. Ideally, the feedback has left your user in a place of confidence. Take advantage of that momentum by sending your user a step further down the path toward their goals.

Video games do feedback especially well. In fact, the quality of the feedback that video games give players is directly related to how much people enjoy the game. Clear and immediate feedback hooks people. This is true even when the game looks like it might be boring to play. Consider this game, *Glider Classic*, in Figure 7.5.

It looks dull, but in studies comparing it to realistic first-person shooter games like *Call of Duty: Advanced Warfare* in Figure 7.6, people found it just as engaging.

FIGURE 7.5

Glider Classic looks like it would be boring to play, but the instant and clear visual feedback hooks players.

FIGURE 7.6

Call of Duty: Advanced Warfare offers a similar quality of immediate, clear visual feedback to *Glider Classic*. The two games are equally interesting once users start playing.

Both *Glider Classic* and *Call of Duty* offer specific, clear, and immediate feedback on player actions that ignite an ongoing engagement cycle. The only difference is that one game gives feedback in the form of a paper airplane's movements, and the other does it with blood spurting from injured opponents.

Let's step through how to plan and implement an effective feedback framework for your users.

> **NOTE** WHAT FEEDBACK DOES
>
> Psychologists think about feedback from a couple of different lenses. Feedback serves an *informational* purpose in the sense that it provides people with metrics by which they can evaluate their behaviors. It also serves an *emotional* purpose in the sense that it can fill people with optimism and self-efficacy, or it can deflate their confidence. Feedback can also serve a *social* purpose. It helps people identify others who are similar to them or may be good models for learning and growth.
>
> Feedback also comes in different flavors. There's *verbal* feedback, which you include in your products via content. There's *visual* feedback, like seeing a burst of light on the screen after playing the right note in *Rock Band*. There's *auditory* feedback, like a buzzer when a game show contestant gives a wrong answer. And there's *haptic* feedback, like the vibration your phone or video game controller might give after an action. Most digital products offer multiple forms of feedback that reinforce and support each other.

Step 1: Measure Something

Before you can give feedback on something, you have to measure it. What you measure is usually a function of a few different considerations:

- The goals your users are working toward (examples might include saving enough money for retirement at age 65 or practicing running until they can finish a 5k)

- The data to help people gauge their progress toward that goal and make meaningful adjustments in their behavior

- The features and capabilities of your product, which can include technical, legal, or regulatory considerations

Once you decide to measure something, it may influence people's behavior even without you providing further direction. I enjoy the "fly in the urinal" example.[1] Apparently, men's restrooms are somewhat unsanitary, thanks to poor aim at the urinals. The cleaning manager at Amsterdam's Schiphol Airport decided to do something about it. He installed stickers inside the bowl of urinals that depicted a fly at rest near the drain, like in Figure 7.7. That's it—no instructions, no cautionary warnings about peeing on the floor. The result? The cleaning manager reports an 8% reduction in cleaning costs, and now this type of sticker is popping up in urinals worldwide (or so I am told).[2]

FIGURE 7.7

A small sticker of a fly is enough of a visual cue to direct more urine down the drain and less onto the floor.

Although the fly doesn't provide feedback in an explicit sense— there's no message saying "Great aim, Timmy!" or "A little to the right!"—it does provide information about where a stream is best directed and a satisfying sense of having achieved the goal or not (again, so I am told).

There are lots of related examples in health, where fitness, diet, and biometric tracking are in vogue. One common measurement in health is step-counting. Tracking activity with a Fitbit, Apple watch, or other device leads people to take about 27% more steps per day than they did before they used it. This occurs without setting a deliberate goal to move more. Just knowing how many steps they were taking gets people to up their movement.

This finding makes sense when you consider what a baseline measurement does. It takes something people had vague ideas about—"I eat pretty healthy," or "I'm active enough"—and makes it specific. Once that happens, they can't ignore the truth as easily

---

1 I first learned about this in the book *Nudge* by Richard Thaler and Cass Sunstein.

2 Similar stickers are also available on Amazon, with heat-activated varieties. They're mostly advertised to parents who are toilet-training their kids, but I can imagine a broader market.

anymore. People who see a discrepancy between what their behavior really is and what they want it to be will be motivated to make adjustments, even if they're small.

Of course, you also need to measure a baseline in order to know how a person's performance changes over time. Collecting baseline data about your users at onboarding will help you give them meaningful feedback as they begin to progress through their behavior change journey.

Measuring a baseline is a critical first step in using feedback to change people's behavior. Next is to offer feedback geared to move people through tasks or stages that get them incrementally closer to a goal.

Step 2: Offer Feedback at Multiple Levels

With any sort of long-term behavior change, people will do better at times and worse at times. Good feedback accounts for normal setbacks by acknowledging them while encouraging forward movement. If someone had a bad day and did really poorly on a test, let them know that's not great, but also remind them that before this they'd had a whole week of better scores. Reminding people of their ongoing success helps them climb back in the saddle by supporting their feelings of self-efficacy.

This pairing of immediate and long-term insights is giving feedback at multiple levels.

The image in Figure 7.8 from the video game *Rock Band* shows what feedback at multiple levels can look like:

- Here you see *immediate feedback* on what the player just did in the form of the flashing lights and colors at the bottom of the screen. If the player hit the right button at the right time, they got positive feedback; if not, they got an indication as to what type of mistake they made so that hopefully they could improve as the song continues.

- You also see *cumulative feedback*—how the player has done over time. This comes in the form of the running total score and power-ups that the player can earn with consistent good performance.

- Finally, there is *normative feedback*, or information about how the player's performance compares to other people's. Here, that takes the form of a leaderboard.

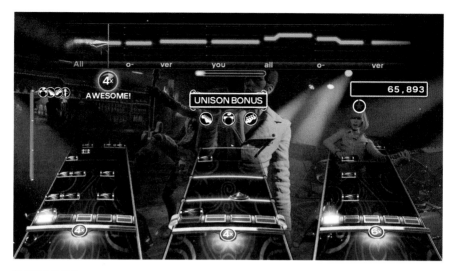

FIGURE 7.8

Rock Band provides users with feedback at multiple levels: immediate, cumulative, and normative.

What's important about having multiple levels of feedback is how each type of feedback provides a different type of information and emotional value to the user. Immediate feedback is helpful for calibrating actions in the moment. It comes very close in time to a behavior so that the information provided can influence a person's understanding of the right behavior. In the *Rock Band* example, giving feedback with a delay of even a few seconds would make it difficult for players to figure out how to improve.

Immediate feedback can evoke different emotional responses, depending on how well a person is doing at the activity. Positive feedback usually makes people feel pretty good, while negative feedback can feel crappy to receive. That's true even if the feedback is about an inconsequential behavior like how well you played a video game. It can also be frustrating to get repeated negative feedback when you've been trying to improve at something, or you expected to be good at it, but it turned out you weren't.[3] And it can be maddening if you've been making progress at something but then have an episode of poor performance (or, as is often the case in behavior change, plateaued).

This is where feedback at multiple levels provides the most value.

---

3 Like me with *Rock Band*.

The Shapa is a scale that doesn't tell you how much you weigh. It doesn't even have numbers on it, as you can see in Figure 7.9 below. Even though it doesn't do the one thing a scale is *supposed* to do, research shows that people lose more weight using it than they do a regular scale, and they're more likely to keep using it for longer.

A lot of people who want to lose weight get frustrated with weighing themselves daily. People's bodies naturally fluctuate by a few pounds from day to day, which can make it hard to tell whether they're really losing or gaining weight. That's especially maddening if they're working hard to see the number go down.

Shapa measures your baseline body composition over a training period and then gives you a color every day that tells you whether your body has significantly changed from your baseline. You'll see green if you're maintaining, black and gray if you're gaining, or blue and teal if you're losing weight, like in Figure 7.10.

With a regular scale, people can't see the pattern in the noise of their data. Shapa's feedback tells users only what they really need to know, and keeps them more engaged as a result.

You went from Green to

Teal

We want to squeal, your color's Teal!

FIGURE 7.10
If you've lost a meaningful amount of weight, the Shapa app will provide you with a teal screen to let you know.

FIGURE 7.9
The Shapa scale has no display to show people their weight.

A lot of what you ask people to do in behavior change takes a long time to accomplish. Consider weight loss. Someone with a lot of weight to lose might be looking at months or years to reach a goal. And realistically, during that time, there will be times when the pounds come off, and times when the person plateaus. There will be times when the person feels sick and can't work out, or they go on vacation and decide to indulge temporarily. It can be incredibly discouraging to receive negative immediate feedback during these times (i.e., when you step on the scale on the first day home from Hawaii). What can make that setback feel better is putting it in the context of overall pounds lost, or inches lost, or physical stamina gained, and offering a strong call to action to get back on track. It could be the difference between someone giving up and someone getting back on the proverbial horse. Providing the bigger picture can help people be more resilient when they receive negative immediate feedback.

Step 3: Add Information About What Other People Do

Normative feedback can also play a critical role in helping someone understand their performance. Normative feedback compares one person to other people. It provides a context that helps people to interpret the quality of their own performance. The leaderboard in *Rock Band* is an example.

This context can help drive behavior change as people realize their own behaviors are relatively better or worse than other people's. For example, you might get one of those reports with your electric bill that tells you how your utility usage compares to the neighborhood as a whole and your "similar neighbors" in particular. Usually, there's a bar graph showing who uses the most electricity among the three groups. There's also an indicator in the form of a smiley (or frowny) face letting you know if you compare positively or not. In Figure 7.11, you can see an example from Opower.

Overall, this feedback seems to prompt meaningful behavior change in people who get the report—they reduce their energy consumption by a few percentage points. Interestingly, this only happens when there's also information that indicates whether the recipient's household is doing well, like a smiley or frowny face. The first iteration of these reports just had the bar graphs. The designers thought that recipients would understand that they should keep up the good behavior of using less electricity than their neighbors. That's not what happened. The people who were doing well started to use *more* electricity once they saw the report. It was like they figured, "I did well in the past so I can treat myself to keeping the lights on in the future." The small praise of the smiley faces fixed this strange behavior overcorrection.

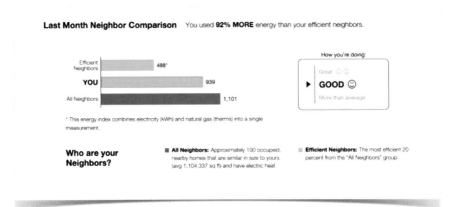

Last Month Neighbor Comparison You used **92% MORE** energy than your efficient neighbors.

Efficient Neighbors 488*

YOU 939

All Neighbors 1,101

How you're doing:

Great ☺☺

▶ **GOOD** ☺

More than average

* This energy index combines electricity (kWh) and natural gas (therms) into a single measurement.

Who are your Neighbors?

■ **All Neighbors:** Approximately 100 occupied, nearby homes that are similar in size to yours (avg 1,104.337 sq ft) and have electric heat

■ **Efficient Neighbors:** The most efficient 20 percent from the "All Neighbors" group

FIGURE 7.11

OPower's feedback shows how a customer's utility usage compares to their entire neighborhood, as well as people who live in similar housing.

So it seems that the most effective normative feedback includes at least two ingredients:

- How your behavior compares to someone else's
- Some indicator about whether that is positive or negative

Another aspect of normative feedback that can be important is who users are being compared with. People respond best to comparisons with people who are similar to themselves, or maybe just a little bit better on the relevant tasks. (In marketing, they refer to these people as *aspirational targets*.) These are realistically people you could be like if you worked a little harder at it—they're 5 pounds lighter, 30 seconds per mile faster, earn a B+ to your B.

Choosing the right comparison group for someone is similar to choosing the optimal level of challenge. If you compare people to a group they're much better than, it's boring. I'm not receiving useful information about my math skills if you compare me to a group of first graders. I *should* beat an average six-year-old on a math test.

Similarly, I won't get much utility from being compared to a group of applied mathematics Ph.D.s. They better outscore me or else find a new career! But if you compare me to other adults with a roughly similar educational background, then I can get a sense of whether my skills are average for my group or not. Within a group of similar people, being a little better feels good, and being a little worse can be a motivator. If people like you can do something, so can you, if you just worked a little harder.[4]

In designing for health behavior change, picking the right comparison group for normative feedback can be tricky. Consider changing the comparison group depending on the variable. Let's take a weight loss challenge. You probably don't want to compare everyone in the challenge to each other on a pound-by-pound basis. Some of them will lose weight much more quickly than others, say because of faster metabolisms. The people with only a few pounds to lose will reach their goal long before the people with lots of weight to lose. So, you could take the slice of people in the challenge who most resemble each other with respect to weight loss goals and compare just within that slice.

You could also pick a different factor besides weight lost to compare them on, especially a psychosocial factor that's less likely to make people feel judged. Examples include:

- You're part of the 20% of participants whose favorite type of exercise is walking.

- Like you, many people trying to lose weight say they sometimes feel self-doubt; you're not alone.

- 55% of the moms in this program say that wanting to be active with their kids is a key reason for participating. Sound familiar?

In these examples, the normative feedback plays less of an informational role with respect to performance, and more of an emotional role in helping build self-efficacy and interest in continuing to participate.

It also may be the case that normative feedback doesn't work for your product for some reason. Not every behavior change journey needs every single type of information loop to be effective. Less may be more.

---

4 This is the growth mindset speaking up again.

Strength in Numbers

Seeing how the individual efforts of many people add up can be hugely motivating to people trying to change behavior. When behavior change goals are focused on societal level change, people's individual contributions to solving big problems can feel paltry, especially if they don't notice any effects in their own life. Consider someone who has chosen to eat less animal protein because they believe it will help sustainability efforts. Cutting meat out of two or three meals a week might not feel like it's making much of a difference. The Darwin Challenge app gives its users evidence that their small changes contribute to a bigger societal shift. In Figure 7.12, the Darwin Challenge shows a tally of the cumulative benefits from all of the plant-based meals its users have logged.

FIGURE 7.12
Users of the Darwin Challenge can see how their own relatively small behavior changes contribute to a much larger impact when combined with everyone else's.

Other examples include MapMyFitness's Year in Review in which they take the cumulative miles run by all of their users and describe the fantastic voyages they would cover. In 2018, MapMyFitness users logged enough miles to run to the moon 2,000 times.

Step 4: Find a Frequency

It's a good idea to let users peek at their progress on their own whenever they're interested, but many digital products will also push feedback to users. The timing of that push feedback is important.

How often should you give feedback on a single goal? Too much feedback can lead to *alert fatigue*. That's when people ignore your messages because they're so frequent or meaningless that they're not worth even a few seconds of attention.[5] And too little feedback can leave people aimless.

There's no magic timetable for the right amount of feedback. It depends on the behaviors you're encouraging and how often they're performed. Some people will prefer more feedback and others might like less. A rule of thumb is that you can give feedback after every *significant* single behavior. If the behavior change is something more cumulative, once a day, week, or even less often may be more appropriate.

Here are some examples of how often you might give feedback for different types of behavior change:

- After every workout, blood sugar measurement, or learning assessment
- At the end of the day for calorie consumption or number of steps walked
- At the end of the week for weight lost or gained
- At the end of the month for money saved or progress made on learning a language

Feedback frequency is a great candidate to test with users during your development process and to see what works for your product.

Praise Be to the Repeat Achiever!

One mistake some behavior change products make is only rewarding someone the first time they do a behavior. If the behavior change

---

5 Coco Chanel famously advised taking off one piece of jewelry before leaving the house to avoid being overdressed. Similarly, consider reducing the number of alerts your product sends users by at least one. If it was really necessary, your data would tell you and you could add it back.

6 For more on designing digital experiences for children, check out Debra Levin Gelman's *Design for Kids: Digital Products for Playing and Learning.*

goal includes repeating a pattern many times, then you want to make sure you can offer praise regularly for people who are consistently achieving the pattern.

Fitbit, which tracks steps and other physical movement, awards its users badges when they hit milestones for the first time. When you open an app after hitting a milestone, you see a celebratory alert with the badge and what you did to earn it. I really enjoyed seeing those alerts! At least, I did for the first week that I owned a Fitbit.

The badges in Figure 7.13 are the ones I earned in my second week of Fitbit ownership, when I carried it in my pocket running a marathon.[7] That's the day I hit 55,888 steps. In addition to the excitement of crossing the finish line, when I opened the Fitbit app on my phone it basically threw a party for me. It was all very fun.

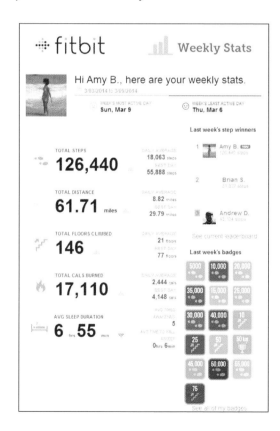

FIGURE 7.13

If I had realized this was the last time I'd see this much Fitbit badge action, I would have savored it more.

---

7 The first rule of running a marathon is that for the rest of your life, you have to tell people you ran a marathon.

And then I never saw another alert in the almost five years I continued to use a Fitbit. They only throw you a pocket party the *first* time you hit a milestone. That means that every regular day I was trying to get my 10,000 steps, I got nothing. The days I got 25 or 30,000 steps, not a peep. There was no way Fitbit was going to praise me unless I ran another marathon, and well, even I'm not that desperate for praise.

To me, this seems like a missed opportunity to spike a regular fitness routine with the occasional exciting digital surprise. Feedback doesn't always have to be about newness. It can be about consistency, too.

Step 5: Tee Up the Next Action

Most behavior change design is about getting people to take ongoing action toward a bigger goal. One-time action won't cut it. So good feedback needs to set people up to take the next step on the behavior change journey. Like Mario in Figure 7.14, your user's princess is in another castle, and your job is to help them get there.

FIGURE 7.14
Poor Mario. Always another castle!

The call to action can be as subtle as showing that the next step is available (like Duolingo does), or as explicit as telling the person exactly what they need to do next.

> **NOTE** GAMIFICATION AND BEHAVIOR CHANGE
>
> You may be wondering where gamification fits into behavior change design. Good gamification directly supports many of the behavior change mechanics in this book, and especially in this chapter.

"Gamification" means the inclusion of game elements in a non-game experience. The result may or may not feel like a game itself. Duolingo, Fitbit, and the Starbucks app are all gamified, but none of them would be mistaken for a true game.

Game elements include points, leaderboards, levels and leveling up, challenges, badges, and rewards. These game elements can all also support effective feedback. In fact, if applied correctly, game elements make an experience engaging because they support people's need for effective and meaningful feedback.

Watch Outs: Common Missteps

One common misstep in setting up a feedback structure is focusing on behaviors that aren't critical for meaningful outcomes. Sometimes this happens because observing and giving feedback on the truly critical behavior is technically difficult or outside the scope of a product. Sometimes people get fixated on a proxy measure as being more meaningful that it is. Here are three examples of how feedback has been tied to the wrong behaviors with negative consequences.

Perverse Incentives

Supposedly this is true, and it's horrifying. When India was under British Colonial rule, the British government was concerned about the number of poisonous snakes in Delhi. Me too. So they decided to pay people by the snake to help eradicate them. What do you think happened?

Aside from the town hall becoming a terrifying snake graveyard, this reward system actually made the problem worse. People began to breed snakes so that there would be more of them to kill and exchange for money. There were so many snakes.

Now when the wrong feedback mechanism ends up exacerbating a problem, it's known as the *cobra effect*.

Incidentally, there's a similar story about the French government's attempt to keep the rat population under control when they ruled Vietnam. It seems that colonial regimes are particularly bad at aligning behaviors with feedback and rewards.

The A Student

As part of healthcare reform in the U.S. under the Affordable Care Act (ACA), hospitals are required to give patients a discharge survey (HCAHPS). The survey scores have implications for how much hospitals are reimbursed by health plans, so getting a high score is important.

One of the questions is about pain management. It was phrased like this:

> During this hospital stay, how often did the hospital staff do everything they could to help you with your pain?

Imagine you're a doctor with a patient struggling with post-operative pain. How do you decide to treat him, knowing that his responses to this question when he's leaving the hospital may impact your job?

As you might imagine, providers tried to maximize their score. They gave medication—the fastest way to reduce pain quickly. This practice may have contributed to an increased use of opioids at a time when addiction was becoming a serious population health problem.[8] Not good!

Gaming the System

A type of feedback common in digital apps is "badging." Badges offer a visual icon tied to a repeated behavior, like logging data or "checking in" to a physical location, and can be a good way to indicate patterns in someone's behavior.

In 2010, Foursquare (now known as *Swarm*) thought to use their badges as a way to encourage physical activity. They partnered with a few fitness apps and created badges that would be awarded if people checked in from gyms or health clubs.

As you may have already guessed, this was not the right behavior to provide feedback on if the goal was encouraging physical fitness. Standing near a gym does not confer six-pack abs, unfortunately.[9]

---

8 The problem grew so pronounced that in 2017, the Centers for Medicare & Medicaid Services decided to remove the pain management question from HCAHPS starting in 2020.

9 For some of us, neither does going inside the gym. Damn genetics.

What Foursquare found was that people gamed the system in order to earn the fitness badges. The top scorers were people who lived or worked really close to fitness facilities and could check in without stepping inside. Foursquare discontinued their fitness program in 2012.

Good Feedback Is for the User

Above all else, the primary purpose of feedback in behavior change design is to encourage growth and progress. It can be tempting to anchor feedback on business goals rather than behavior change goals. And sometimes it's appropriate to use feedback to nudge users to upgrade to more robust functionality or purchase an additional product, if that's the course of action that's genuinely most likely to help them toward their goal.

But users are pretty savvy at sniffing out when feedback is about them and when it's about you. Keeping people engaged requires keeping most of the focus on *their* needs. Fortunately, if your product offers people effective behavior change support, you shouldn't need to go overboard with the sales pitch.

The Upshot: Feedback for All

Don't be shy about offering your users feedback early and often. It's one of the most powerful mechanisms you have to help people understand how their behavior change efforts are going, and make the necessary course corrections. If you understand how to craft good feedback, you can use it to help your users persist through setbacks, "level up" their skills, and remain engaged with behavior change for the long haul.

There is an art as well as a science to feedback, but you will be well-served by five basic steps. Measure something. Give feedback at multiple levels. Include information about what other people do. Schedule the feedback to arrive when it can be most helpful. And guide your users to the next step.

Workplace performance is a fertile area for behavior change design, but it can be challenging for behavior change designers to find examples of well- and poorly-crafted performance management tools because they're usually accessible via a B2B arrangement. I talked with Diana Deibel of Grand Studio about her experience redesigning an employee performance management tool that wasn't working well, and how she used behavior change design tactics, especially ones related to feedback, to improve it. Diana's case study shows how the behavior change design principles used for health, financial, and educational behavior changes can also work to support workplace performance.

What wasn't working?

At Grand Studio, we have something called *culture week*, which includes introspection and self-analysis workshops. We were using a proprietary performance management tool to track employee skills and growth that was designed to be flexible and something everyone could use, no matter what their job title or level was. One of the things that became apparent out of the culture week discussions was that it was both too vague and too overwhelming (Figure 7.15).

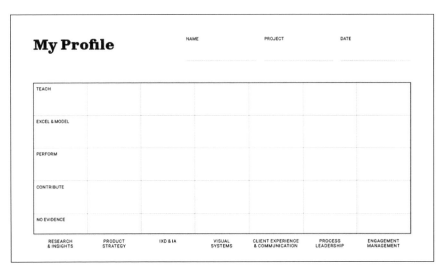

FIGURE 7.15

The original performance evaluation dashboard for Grand Studio employees was designed to be flexible, but actually ended up being difficult to use because of a lack of specificity.

How did you make it easier to understand what success looked like?

Our team completely redesigned the performance management tool (see Figure 7.16). We wanted it to be clearer and more attainable to both the employees and the managers using it. Rather than having one tool that was used across levels, we created one for each level of designer we currently employed (Designer, Senior Designer, Lead Designer) that had metrics applicable to what was expected of that role and a checklist of things they needed to learn in order to get to their next level. Each person still had the autonomy to choose the core skill in which they wanted to focus (research, visuals, UX), but they had an equally weighted and specific checklist geared toward their achievement of the next level of their career.

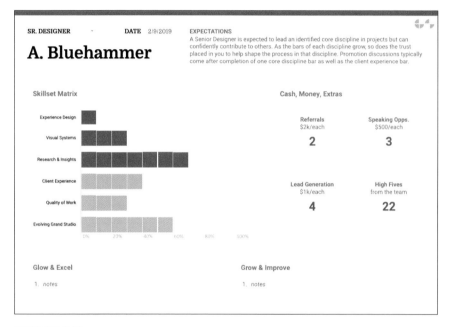

FIGURE 7.16

The redesigned performance evaluation dashboard sets out specific core job responsibilities while also giving employees the flexibility to choose growth areas—and be financially rewarded for them.

We wanted to break work performance into small attainable steps toward goals and help employees track their progress on achievement. To do that, we created a dashboard that included progress bars which filled up a bit more with each item checked off a list of steps. We also included positive and critical feedback from peers, along with an action plan with specific goals to check off their next box. This type of 360 feedback paired with actionable steps gave employees a sense of their strengths/weaknesses along with a way to have tangible agency in their own career. We listed the positive feedback first to reinforce the feeling of accomplishment and capability and create more receptivity toward the critical feedback.

How can you give people a feeling of control in an employment context?

Beneath all the "countable" metrics and peer feedback, we also included ancillary metrics, which were entirely optional and held no sway over someone's ability to get promoted. These metrics included speaking engagements, hiring referrals, and generating sales leads. We all knew that part of a promotion was recognition and validation, but another part was financial. The additional metrics were financial levers in the employee's control, which they could pull at any moment they felt it was required. Additionally, these badges allowed for positive engagement and active involvement in moments where employees might typically disengage due to lack of opportunity. Working on these tasks also helped prime junior employees for skills they would need when reaching higher leadership roles that included people management, marketing, and business development responsibilities.

Diana Deibel is a lead designer at Grand Studio in Chicago, where she designs human-centered products, focusing on voice user interfaces and the UX research that goes into products people like to use.

CHAPTER 8

Come Together

Design for Connection

There are individual differences in how much people crave connectedness with others. Some people are social butterflies who have large networks of friends and thrive on lots of togetherness, while other people cultivate smaller, tight-knit groups of confidants and need time alone to recharge their batteries (and many flavors in between). However, no one is truly a lone wolf. The human brain is wired for connection.[1] Study after study links positive social relationships to dramatically better physical and mental health. And without relationships, humans don't do well; social isolation is predictive of mental and physical illness and earlier death. People need people.

In self-determination theory, feeling part of something larger than oneself, or relatedness, is a core psychological need for everyone. Social relationships—real or virtual—can help fulfill this need. Behavior change designers have opportunities to weave social connections into their product experiences to help users reap the benefits of relatedness. This chapter covers why and how to do it.

What Can Social Support Do?

Like so much else in life, behavior change is better with friends. Research shows that having companions on a quest for change increases people's odds of success. Better yet, having social support for a behavior change goal influences how much people enjoy the experience of getting there. And when people enjoy something, they're more likely to stick with it, so friends are an ingredient in long-lasting behavior change.

Social connections provide a variety of practical and psychological benefits. Other people can lend a hand to solve a challenging problem, offer moral support or advice for someone who needs a boost, or provide a positive role model to emulate. And many studies prove that being around people they like and care about makes people happier. In the context of behavior change specifically, some of the most important functions of social support include strengthening people's purposes, offering them concrete assistance with achieving goals,

---

1 So are the brains of other primates. If you've taken an introductory psychology course, you might remember Harry Harlow's 1965 study about socially isolated baby monkeys. These poor babies were so hungry for affection and attention that they clung to inanimate mannequins put into their cages. If they spent long enough without social contact with other monkeys, the monkeys never developed normal social skills. It's a reminder that social contact is important for healthy development at the most basic levels.

providing teaching and accountability, and helping people occupy the right head space to be productive with behavior change activities.

Foster Purpose and Fun

In previous chapters, you learned that having a deeper purpose can help people invest in the behavior change process and close some of the gaps between their present day and future selves. Social connections can help people find that deeper purpose in their activities, whether it's because the social part itself has meaning, because people help each other think about the behavior differently, or because it's simply more fun to do things with friends. For example, people who exercise together show up to their sessions not just for the benefits of sweating it out, but also because that time enhances their relationship with each other and helps them focus on something other than burning muscles. They also may develop a new identity related to the group; they're no longer just a person who exercises, they're a Heartbreaker.[2] The social element gives people more reasons to stick with their new behaviors.

People don't necessarily have to share the same goals in order to engage in purposeful activity together. Consider the people who use Busuu, a language learning app. The app pairs native speakers with people learning a language, so they can benefit from authentic accents, slang, and syntax (see Figure 8.1).

FIGURE 8.1
Busuu matches language learners with native speakers; the social aspect of learning a language this way can reinforce the sense of purpose for users.

---

2 I've seen this happen over and over with friends who've joined the Boston-based Heartbreak Hill Run Club.

Within any given pair of users, one has the goal of learning and the other has the goal of teaching. The experience is likely to be purposeful for both.

Consider finding ways to connect people as a part of your behavior change intervention. As a bonus, provide them with specific roles to play that uniquely contribute to each other's progress toward a goal.

Help a Pal Out

Another way people can help each other with behavior change is through *instrumental support*—practical assistance. Instrumental support could look like:

- A loan or financial help to help someone pay tuition and learn a new trade

- Sharing budgeting workbooks that describe how to save for retirement

- Babysitting the kids for an afternoon so the parents can train for a race

Most of the ways that people help solve someone else's problem through action come under the heading of instrumental support. Behavior change interventions can facilitate instrumental support by making it easy for people to ask for or find it, and offering a route for others to give it. For example, MyHealthTeams makes condition-specific social communities where patients and caregivers with a specific condition can do all of the usual social networking stuff but also swap tips about treatments. The provider directory on each MyHealthTeams site is crowdsourced so that the doctors included carry the implicit recommendation of other patients with the health condition (see Figure 8.2).

In the finance realm, Honeyfi makes it easier for couples to manage their joint finances (Figure 8.3). By allowing them to share financial information in a centralized place, Honeyfi enables instrumental support related to money goals.

FIGURE 8.2
The provider directory on MyEndometriosisTeam was developed based on user suggestions, which implies patient recommendations and gives new patients helpful clues about which doctors to try.

FIGURE 8.3
Honeyfi helps couples provide each other with instrumental support managing their shared finances.

Show the Way

People can also support each other in behavior change as mentors. Mentors can and often do provide instrumental support, but their impact goes well beyond that. They offer role models for people to pattern their own behaviors against. They can provide emotional support, coaching people through the feelings that come with hitting a road block or reaching a milestone. And just the presence of another person may cause someone to work harder or perform better at some types of tasks, as the creators of the Fabulous app know (see Figure 8.4).

Mentorship relationships can help both the mentor and the mentee thrive. In one study, researchers trained a small group of patients with multiple sclerosis (MS) to support less experienced patients. The support training specifically focused on active listening rather than providing advice or direction; the patient supporters were, in effect, trained to be great

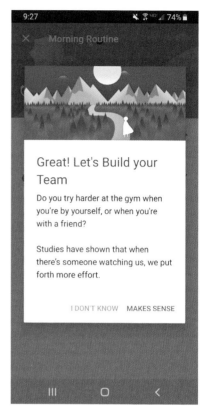

FIGURE 8.4
As the Fabulous app points out, some types of behaviors are performed better when there's an audience. Just having a buddy nearby can make workouts more effective.

listeners. Over the three years of the study, the patients receiving this peer support did pretty well and showed modest quality of life improvements. The peer supporters, however, blew them away with an almost eight times greater increase in their quality of life over the same time period. Helping others seems to have given the peer supporters something vital to their well-being.

It's worth emphasizing that the peer supporters in that study did not have any special subject matter expertise or unusual experience. Active listening can be done by anyone with basic instructions and some practice. When you're considering how to add social support to a behavior change intervention, think about whether there are opportunities not just for users to receive help, but to *give* it. Can you equip them with simple tools, like better listening skills or cheerful memes to share, that they can use to help others? For example, Wisdo pairs users with buddies who have similar experience profiles for two-way support (see Figure 8.5).

Finally, other people offer opportunities for social learning. Recall from Chapter 7 that people who are just a little bit better at something can help motivate people to step up their own game. Those *aspirational targets* are similar enough to your users that with a little effort, they might be able to catch up their performance. Exposing your users to people similar to themselves who have succeeded at some aspect of behavior change can help them recognize new tactics to incorporate into their own routines. Including examples, stories, and normative feedback in your product doesn't just add interest—it may inspire people.

FIGURE 8.5
Wisdo pairs users with each other for support with shared experiences; both users in a buddy pair have the opportunity to help and be helped.

Keep It Honest

People are more likely to follow through on their plans if they know someone else is paying attention. You can help your users identify other people in their lives to serve as "accountability buddies," either formally or informally. A formal accountability buddy knows the role they're playing; the user may have asked them to check in to see how the plan worked out, or promised to report back on the results. stickK is a goal-setting app that encourages users to identify a formal accountability buddy to help increase their odds of achieving goals (see Figure 8.6). Users can opt to go it alone, but stickK makes it clear that's not their best choice.

If users have identified a mentor or coach to support them, that person can also provide accountability. In fact, a mentor may have strong persuasive pull. Once someone is in a position of guidance, the mentee typically doesn't want to let that person down. A desire to follow through for the mentor's sake might motivate people to stick with behavior change beyond where they might have quit if they were going it alone.

Informal accountability buddies are people who may not realize the purpose they're serving in behavior change. They're somehow made aware of the person's goals or progress—perhaps when the person shares something on social media, which apps like MapMyRun make easy to do (see Figure 8.7). Even if they don't specifically ask about how the activity went, knowing that other people can see what they're doing can help your users with follow through. Note, though, that if these posts are too generic, others may not come through with supportive responses, so it's important that users have a chance to add their own commentary to the post.

Here's another fun way to build accountability into behavior change using competition. One study focused on increasing physical activity had people

FIGURE 8.6

The goal-setting app stickK facilitates the identification of an accountability buddy to help users stick to their plans.

FIGURE 8.7

MapMyRun is one of many apps that make it easy for users to share their activities on social media, which can support accountability and follow through.

compete in step challenges with their families. Each day, one family member was selected at random to have their step count be the score for the whole family. Suddenly, slacking on steps could have consequences for the team. What do you think happened?

Who You Are Matters

Everyone has a number of what psychologists call *social identities*. Social identities are based on the groups that people belong to. In my case, my social identities include *woman*, *psychologist*, and *Bostonian*. Most social identities carry some meaning that probably doesn't overlap 100% with how any given person experiences it. That's because the meanings of social identities are partially shaped by culture, so they may include stereotypes or overgeneralizations.

When people become very aware of one of their social identities, it can have an outsized influence on how they act in a situation. Heightened awareness might happen when someone spends time reflecting on what it means to have a particular identity, perhaps sparked by reading a news story about someone else with that identity. Situations where an identity is emphasized can also make people ultra-aware of it. Being the only woman in a boardroom with men will heighten a woman's female identity, as will attending a women's leadership conference where there are no men at all.

In digital behavior change interventions, one tactic to subtly activate social identities is image tailoring. It requires you to know demographic information about each user and be able to personalize the images in the program based on that data. If you have those capabilities, pepper your product with photos of people who share demographic characteristics such as (rough) age, sex, and ethnic background to your users. You can include other cultural signifiers, too, like the Dutch researchers who created different versions of a fruit- and vegetable-eating program for ethnically Dutch, Turkish, and Moroccan women that included colors and patterns associated with each culture (Figure 8.8). Culturally relevant visuals will activate the social identity. As a bonus, research shows that people are more likely to pay attention and remember information when it's presented alongside pictures of people who look like them.

FIGURE 8.8
Depending on their ethnic origins, women participating in this Dutch program encouraging fruit and vegetable consumption saw a different visual treatment of the information.

When people are reminded of a social identity, its underlying meanings can affect how they behave. Often, people don't deliberately choose to do anything differently because of their social identity; they may not even be aware they *are* doing anything differently. For example, in research studies, Asian American women's math scores were higher when they thought about being Asian, but lower when they thought about being women. The scores became consistent with the stereotypes associated with each of those identities.[3] The differences are small, but when you're thinking about maximizing your SAT score, even a few points can make a difference. People don't need to believe that the stereotypes are true to have these performance changes, either. Being exposed to them over a lifetime is enough for them to have an influence.

Social identities can be a resource for people to explicitly marshal to help them behave a certain way. Have you ever found yourself gracefully handling a difficult conversation with an angry customer when a similar conversation with your partner would lead to a screaming match? You were probably able to draw behavioral guidance from your professional identity that helped you squelch your irritation.

---

3 These findings don't hold true in parts of the world where stereotypes about Asian or female academic performance are different, which suggests it really is the cultural myths around social identities that influence behavior. Recent replication attempts have only worked when the people in the study were familiar with the stereotypes about the identities.

You may not always know what social identities your users have. Not to mention, it can be tricky to suggest that people think about a social identity if you don't have a firm understanding of what it means to them. Fortunately, you can fake the performance effects of pre-existing social identities by creating new ones within your product. That's what DietBet did when it informed its users that 96% of "dietbetters" lost weight (see Figure 8.9). Even though the identity of "dietbetter" was meaningless before the user joined the DietBet program, the context made its meaning clear. So when a DietBet user thought about cheating on their weight loss plan, they could think about whether that was really something a dietbetter would do. What new identities can you assign your users to help guide behavior?

Will this new identity of "dietbetter" have meaning as nuanced and long-lasting as other social identities? Almost certainly not. Will it provide strong enough guidance to get people through rough spots during a short-term behavior change program? Perhaps.

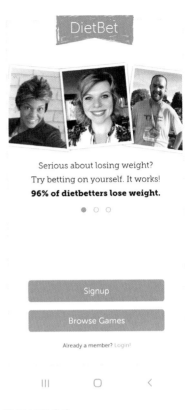

FIGURE 8.9
DietBet calls its users "dietbetters" and reminds them that most dietbetters are successful at losing weight. This creates a new social identity for them.

And Just Be Authentic

It's easier said than done, but "just be yourself" is pretty good advice. One of the oldest theories of motivation in psychology is Maslow's hierarchy of needs (see Figure 8.10). The hierarchy describes a kind of rank order of the things that people need to be happy. The most basic physical needs, like food and water, are at the bottom of the pyramid. The needs get increasingly complex and psychological up the pyramid, with the top being "self-actualization"—living into one's full potential. Another way to think of self-actualization is full authenticity. By giving people opportunities to express their authentic selves, behavior change can help make people happier.

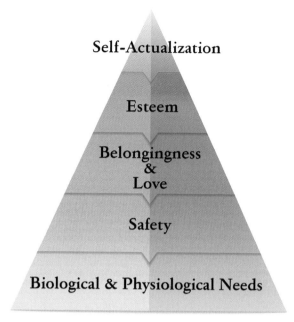

FIGURE 8.10
Maslow's hierarchy
of needs has self-
actualization, which
includes authenticity,
as its pinnacle.

Not being authentic requires emotional labor. Emotional labor refers to having to manage emotions and feelings in order to fulfill requirements for a role. It's a huge component of many people's jobs, especially if they have to provide customer service. If you've ever had to do emotional labor for any period of time, you know how exhausting it can be.

One of the key benefits of being with people who know and accept you is being able to put that work aside and not spend the extra energy. In Chapter 4, you learned that people can improve their decision-making by avoiding unnecessary stresses. Given the demands of behavior change, authenticity can help relieve the stress of putting on a front while freeing up mental resources to devote to new routines and good decisions.

When people are able to be authentic, it also opens up the opportunity for their real selves to be seen and regarded positively. Research shows that people are good authenticity detectors, and respond more positively to people who are real. Since relatedness is all about that deep type of connection, facilitating authenticity can make a big difference in the quality of social support. In fact, when people interact online as part of a weight loss program, the ones who are there for their own reasons (as opposed to being paid to post to stimulate discussion) tend to post more motivational content.

Note that being authentic does *not* mean acting without a filter. Depending on the hat you're wearing at a given moment, authenticity looks different. You probably don't act exactly the same at work as you do at home or as you do at the bar with friends (unless you have an awesome job or a terrible local pub). But Work You and Weekend You share some core characteristics that you express a bit differently because of the demands of your role and the situation. Research shows that even though a person might behave differently in different contexts, they can still reap the benefits of authenticity in each situation as long as those core characteristics are expressed. The trick is being *the right version of yourself* for the situation.

> **TIP** AUTHENTIC GOALS
>
> Authenticity isn't just important for social support. Like you learned in Chapter 3, it's also critical that people's behavior change goals align with their genuine desires if the change is to be long-lasting. Give your users opportunities to set their own goals and frame their goals in a way that makes sense for their lives. They may choose *aspirational goals* that speak to their imaginations, but with your assistance these can be sculpted into concrete and achievable steps.

Social Support in Real Life

Behavior change design can help fulfill people's need for connection by enabling their relationships with other human beings. That may look like helping users talk to their existing social connections about their behavior change journeys or facilitating new connections specifically to support behavior change. In connecting people with one another, it's important to be mindful of both *who* is being connected and *what* they can offer to the process of behavior change. Some pairings, like chocolate and peanut butter, result in delicious success; others, like pickles and ice cream, probably won't work for your average user. Here are some considerations for successful behavior change matchmaking.

Ask the Right People...

Sometimes users need coaching and guidance to ask their network for help in real life. Try to help users recognize the people in their life who will be excellent partners or cheerleaders for behavior change. Like the MS patients who were trained as peer supporters, these people don't need to have any special skills, just a willingness to listen and help if asked. A good rule of thumb is that people who are already doing the target behavior regularly in their own life are more likely to be good supporters for someone working on that behavior. If someone is trying to curb their credit card debt, they'll likely have an easier time hanging out with a frugal friend than a high roller one.

Unfortunately, not all friends are equally supportive when it comes to behavior change efforts. It's not always easy watching someone change, especially if they're doing something that you might wish you could do yourself, or if the change means your relationship will also shift. Someone who's struggling with their own feelings of jealousy or loss might not have the emotional capacity to be really supportive. In some cases, they may even lash out or try to sabotage the behavior change. People who try to cobble together their support network from these obstructive types may be worse off than if they'd tried to go it entirely alone.

Your users' extended social connections matter, too. Research shows that people are more likely to engage in a given behavior like smoking or eating poorly if their friends' friends' friends do. There are probably several reasons for this effect. People gravitate toward others like them, so smokers may find themselves making friends with other smokers more easily than nonsmokers. And when a lot of people you know do something, it normalizes the behavior and makes it easier for you to do without feeling bad about it. The upshot for behavior change is that these sorts of extended social networks matter.

Part of your intervention may include helping your users identify people in their lives who are either helpful or unhelpful for behavior change. That could include having them make lists of people who fit certain categories (like "is a morning person I can call before work if I need to talk" or "will give it to me straight if they see me cheating"). It could also include seeking out new relationships to fill gaps in their support system—something that Sober Grid facilitates for its users who are trying to stop drinking alcohol (see Figure 8.11). That doesn't necessarily mean making new friends (although that could be a nice side benefit).

People can create supportive interactions with other people to supplement the more enduring relationships they have with friends or family. There's a concept called "high-quality connections," referring to any pairing of two people where they both have a positive subjective experience. Research has shown that these high-quality connections, whether they last just minutes or endure for years, benefit both people.

There are five steps that people can use to transform an interaction into a high-quality connection. They are:

- Being present, i.e., paying attention and focusing on the moment

- Being genuine or authentic

- Communicating affirmations by pointing out positive characteristics of the other person or praising something good the person has done

- Effective listening that includes listening for real or deeper meaning

- Offering support in communications, which includes making requests instead of demands and offering criticism that centers on behavior and not negative qualities of the person

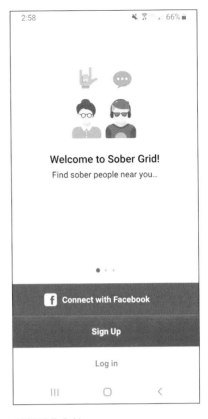

FIGURE 8.11
Sober Grid helps users identify other people nearby who are also sober, so they can spend time with people without feeling pressure to drink.

These tactics can be applied within any relationship. Your users are more likely to get encouragement from each other if they've interacted on an online message board or social media platform using these tips.

For the Right Things

Whether it's asking someone for instrumental support of some kind, trying to enlist a friend to be a partner in behavior change, or simply seeking a listening ear, people often need help to recruit others effectively. Behavior change designers can offer that.

A first step is helping users craft their request for support. People aren't sure how to approach the conversation where they ask someone else for help related to behavior change. Not everyone is comfortable asking people for support. And people who do feel comfortable making the request may not have the skills to do it effectively. A vague request like "Will you support me while I try to lose weight?" isn't likely to yield meaningful help. The recipient of the request may want to be helpful, but be unsure how to do so. They may end up doing nothing, or worse, doing the wrong thing. Behavior change interventions can help coach users through planning a conversation where they ask someone else for support.

> **TIP** MAKING THE ASK
>
> The most effective requests for help are specific about what help looks like. Depending on the focus of your product, you may want to provide users with a list of helpful things to ask for, based on what you know helps others like them. Or you can have users generate their own lists of what type of help they need. Then they can insert those items into their requests: "Would you be willing to babysit for an hour before our movie date so I can squeeze in a run?" A specific request like this is far more likely to generate a positive result than a vague one.

After users have shaped the content of their request for help, it's time to actually present it. People who are reluctant to ask for help that they need may benefit from tools that facilitate the conversation. Some apps will let users send requests to other people to either join them in the app, or to provide some form of support outside of the app. Sometimes those requests double as recruitment tools for the app, like with Gyroscope (Figure 8.12). This particular request makes the mistake of emphasizing the financial benefit of referring new users over the possibility of helping friends. The latter is more likely to spur action. It's easier to ask a friend to do something because it will help them, rather than because it will earn you a small discount on a subscription.

A better way to facilitate user outreach for support comes from SuperBetter. Users of the SuperBetter app are encouraged to ask people in their real lives to be allies in a behavior change quest (see Figure 8.13). They can choose to use the lens of the SuperBetter game if their intended ally has a playful spirit; if that doesn't strike the right note, they can ask for help without revealing the game.

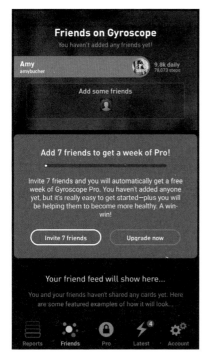

Other times, people prefer to connect with others specifically for
the purpose of behavior change. They may feel like their goals are
too private to share with people they know in real life. If the target
behavior involves a health condition, injury, or other uncommon factor, people may prefer that their behavior change buddies also belong
to that group. Again, behavior change design can help match interested people to each other specifically for the purposes of supporting
each other through change.

Supportive Combat

Social support doesn't necessarily have to be one person encouraging another one toward a goal. It can also take the form of people
working together toward a shared goal, in which case they may work
as a team, or compete against one another to reach the goal first.

When this competition is mutually agreeable and the competitors are evenly matched, a behavior change rivalry can be a source of fun and motivation. The first rule of thumb for competition, then, is to either sort users into groups of approximately equal ability or offer them the opportunity to forgo competition altogether. Don't pit a rookie against an expert!

Direct competition can also be tricky because it usually involves losing at least occasionally, even when the competitors are well-matched. A lot of people starting behavior changes have low confidence or have experienced past failure, so even a symbolic loss might shake them. One way to lessen the impact of inevitable losses is to have competitions turn over quickly. For example, Fitbit's users can opt into a work-week step challenge that lasts just five days (see Figure 8.14). If someone doesn't win, they can try again next week.

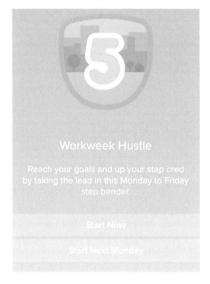

FIGURE 8.14

Fitbit users can compete in short five-day competitions for the most steps.

Finally, the safest competitor for your user is... your user. When people compete against their own historical performance and strive for a personal best, they get some of the benefits of social competition without the bitterness that can come from losing to someone else.[4] Best of all, users will be able to carry on competing with themselves long after their experience with your product is over. Coaching people how to best their own past performance can help with lifelong engagement.

Taking Social Support Online

When the internet first became popular for regular people in the 1990s, it set off debates about what being able to talk online would do to relationships in real life. Would people stop talking to one another? Would they seek solace on the screen instead of face-to-face?

---

4 The video game *Mario Kart* uses this strategy by letting players race "ghosts" representing their own performance in past matches.

Like with so many doomsday debates around new technology, the effects of the internet on relationships seem to be a mixed bag. Yes, people can use the anonymity of screen names to say and do things they'd never do in person—but that cuts both ways. There are the trolls who use their anonymity to be cruel, but there are also people who find the courage to share stories or express themselves creatively when they don't have to worry about being identified. The positive aspects of assuming an online persona could help some people work through difficult issues.

And for many people, social networking enriches their lives by reconnecting them with long-lost friends or allowing them to remain connected over distance. It can also help create new real-world connections that become true friendships. Behavior change design can take advantage of that. Your users may be getting a good deal of their social support through social media or other digitally mediated mechanisms. Contrary to the rumors, social networking is not evil. People can and do successfully use it to knit together effective, helpful communities that support behavior change goals.

Finally, social networking can connect users to experts whom they might not otherwise be able to access. This greatly enhances the resources available to people via their social support systems.

Real-World Social Networking

The first way to use social networking as part of your behavior change design is to facilitate communication between your users and the people who are already part of their networks. This can take the form of online sharing for advice, feedback, and accountability, or it could help foster offline activities related to the behavior change (like taking a Meatless Monday challenge as part of an Earth Day campaign).

One way many behavior change apps leverage social networks is by allowing (sometimes encouraging) users to auto-post messages about their activity on Facebook or Twitter. It turns out, these sorts of messages don't generate engagement. People rarely "like" or comment on them, and so they just create digital spam. Worse, users may mistakenly believe the lack of engagement means that others don't support them, when, in fact, it's more likely that people don't engage with boring content. If you include the ability for people to post about their activity in your program on their networks, be sure to let them customize those posts.

It's important not to make social sharing a requirement of using your intervention. If people feel forced to share information they don't want to, they'll disengage, and nobody wins. Remember that being able to make meaningful choices is a key ingredient in lasting motivation. Being forced to tell everyone in your Facebook network that you're trying to lose weight feels like you're being deprived of a pretty meaningful choice. Letting people customize both the audiences and content for their posts will help restore some of that sense of autonomy.

Networking for Change

Social networking makes it much easier to find people to encourage or participate in behavior change with you in a way that aligns with your preferences—an important part of weaving authenticity into your process. Social networking in behavior change doesn't have to leverage real-life connections. You will have users who, for whatever reason, prefer not to bring their real-life network into their personal goals.[5] Or you may want to help connect users with other people who share something specific in common with them. Programs like Noom (Figure 8.15) can create virtual support groups for people who are starting their behavior change around the same time, so they can work through milestones together.

Many digital behavior change interventions include their own

Finally, in 2 weeks you'll also be placed into a Support Group of people like you who will share your journey through the ups and downs of weight loss.

We'll use this to hand pick groupmates who will be at a similar point in their weight loss journey and able to relate to your struggles and successes.

Next

FIGURE 8.15
Noom connects users who began their weight loss process around the same time in a virtual support group.

---

5 This is especially likely if your behavior change focus area is highly private or stigmatized. Many people will prefer not to widely share financial and health behavior change goals.

homegrown networking capabilities. Others use tools on other social networks to facilitate behavior change conversations among their users—say by creating a Facebook group or a Twitter hashtag.

> **TIP PICKING A PLATFORM**
>
> Unsure which social media platform(s) are the best fit for your product? Ask a handful of people who are similar to your target users where they spend their time online. And remember that you have to have a very compelling value proposition to get people to adopt a new network or tool. For most designers, leveraging existing tools will make much more sense than developing their own.

Research suggests that behavior change–specific social networking can help people work on target behaviors, with some parameters. First, it works best if the people in the network are there because they want to be. Paid participants tend not to elicit the same caliber of response from other members (it's that authenticity thing again). This suggests that behavior change designers should not force their users to participate in social aspects of their product (more on that later).

Second, certain types of support messages are more helpful than others. Dr. Sherry Pagoto and her team found that the types of messages people posted in a Facebook weight loss group predicted their success. The people who lost more weight tended to share their goals or specific accomplishments. They also asked for help rather than just describing their challenges. Behavior change designers might then consider structuring any sort of social networking activities for their users around generating accountability and asking for specific support.

Avoiding Social Media Traps

Social media does have a dark side. Most people portray a more optimistic version of their lives on social media than the reality. People post selfies where they have good hair by the beach, not ones with bleary eyes by the desk. Even though most people edit their lives for social media, it can be hard to realize that other people do it, too.[6] When your users are online seeing other people live better lives than they do, it can complicate their efforts at behavior change.

---

6 This is partially due to a cognitive bias known as the *actor-observer bias*, where people think their own behaviors are due to the situation, but other people's behaviors are due to their personalities.

One maladaptive phenomenon is the *fear of missing out*, abbreviated as FOMO. FOMO is more likely to happen when people's social needs aren't being met; without meaningful relationships to help them feel a sense of community, people look with envy on others' good times. Research has shown that people who have high levels of FOMO are more likely to seek out social media in order to feel connected. Unfortunately, social media can then exacerbate FOMO because so many people use it to share an overly positive version of their lives. Thus, it becomes a cycle of loneliness.

FOMO can also be used to nudge people into a behavior, for better or worse. If your users see testimonials from other people who have enjoyed their experience or achieved strong results with behavior change, it might prompt them to engage more themselves. That's part of what Curable is trying to do in sharing the "unfiltered stories" of other users who have reached their goals of curing back pain through the app (Figure 8.16). A more heavy-handed version of using FOMO to get people to do something is the "Act now!" showmanship on infomercials. It may seem cheesy, but it's common because it often works.

Another problem is that social networking can make people feel even lonelier. When people go online to connect with friends or make new friends, social networking helps them feel more connected. The danger is when people withdraw from their real relationships and seek solace in social networking instead. Worse yet is if they believe the misleading edited versions of people's lives that they post, and see how their own experience doesn't stack up. That hurts.

What's a behavior change designer to do? First, you can help users to put parameters

⚙ Recovery Stories   ✕

TRUE STORIES OF
RECOVERY

Think you're alone in this journey? Think again. These are the unfiltered stories of chronic pain sufferers who were desperate in their search for relief. After multiple doctors, failed treatments, and thousands of dollars spent, they finally found their way to an approach that worked. Hear about their symptoms, struggles, and life-changing journeys to lasting health here.

FIGURE 8.16

Got back pain? Don't miss out on the "life changing" experience of using Curable to address it.

around their social media use so that it's more need supportive. One option is a program like BlockSite (see Figure 8.17), which lets users designate certain sites and programs they're not allowed to use during a scheduled period (or all the time). Second, if you include social networking as part of your product, try to position it as a way to form meaningful relationships rather than a means to its own end. Prompt people to connect with others and give them ideas about what to do with those connections: share stories, ask for advice, even meet up for an in-person workout class or volunteer session. Finally, remind your users when you can that social media is not real life. Even people whose online presence is authentic often protect themselves with selective sharing.

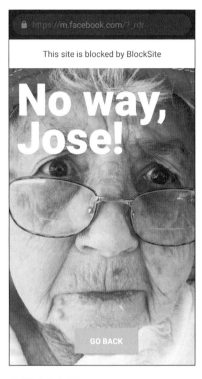

FIGURE 8.17

BlockSite lets users restrict their own access to certain programs at designated times to prevent mindless social media use.

Lurkers Welcome

Vicarious participation in social support can also help people cope with behavior change. BecomeAnEx is an online smoking cessation program that includes a social forum, the EX community (see Figure 8.18). BecomeAnEx is the brainchild of the Truth Initiative, a nonprofit public health foundation that rigorously studies the efficacy of their own products. In one study, they looked at a sample of their users and divided them into three groups: people who never used the community, people who lurked in the community (that is, read the content but never posted any messages themselves), and people who actively participated in the community. Then they compared how many people in each group were still smoke-free six months after their quit date. Perhaps predictably, the people who were active community members had the best quit rate, over 20%, while people who never participated at all had the worst at about 4%. Surprisingly, the lurkers were hot on the tails of the active community participants, with a quit rate of 15%—more than three times better than people who ignored the community.

It's not clear *why* lurking has such a dramatic relationship with successfully quitting smoking. It could be that by reading the advice from community "elders," lurkers gained useful tips or learned what to anticipate as their body got used to a nicotine-free existence. It's also very likely that lurkers experienced a sense of belonging with members of the community, even though they weren't actively participating. There are predictable symptoms that happen when people quit smoking. Even if they had absolutely nothing else in common, lurkers were almost certainly able to read the stories of other smokers in the community and recognize their own physical and emotional experiences of weaning off tobacco. That's vicarious relatedness.

The design takeaway is that if your product includes a forum or other social feature, make it easy for users who don't want to directly participate to lurk instead. Requiring people to actively participate in order to view the social aspects of a product may lead some to abandon the features entirely. So let them lurk and reap the benefits.

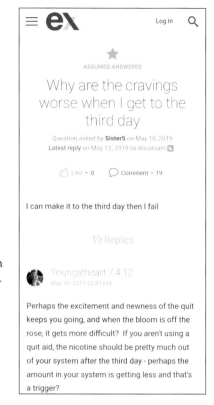

FIGURE 8.18

Participating or even lurking in the BecomeAnEx community forum significantly improves users' chances of staying off cigarettes three months after their quit date.

Virtual Experts

Sometimes people need expert help with their behavior change challenges. Fortunately, digital products can also connect users with live experts to help. Services that provide on-demand access to professionals can offer users an extra boost to overcome obstacles—although it's worth noting that products that include this level of high-touch support often come at a high cost.

Concierge medicine and virtual doctor's visits are one example of how technology can connect people with professionals for guidance. Companies like Teladoc and Doctor on Demand allow people to consult remotely using video chat with physicians for quick input on nonemergency health issues. Meanwhile, Biogen's Aby app to support people with MS allows them to chat with a nurse educator specializing in their condition, as seen in Figure 8.19.

There are programs to help people receive mental health services via digital channels, too. Access to talk therapy can be limited depending on where a person lives and what kind of behavioral health insurance coverage they have. Many people are also reluctant to seek mental health care because of the stigma still associated with it in many circles. Apps like Talkspace let people connect with a therapist for remote talk therapy sessions (see Figure 8.20). In this case, users benefit from both the convenience of a virtual visit and the privacy of an online session.

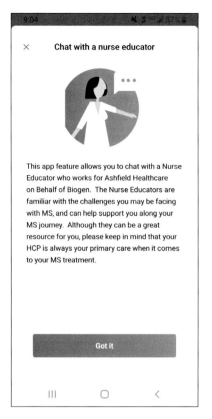

FIGURE 8.19
With its Aby app, Biogen lets patients with multiple sclerosis connect with nurse educators specializing in their condition.

In the fitness realm, the Aaptiv app lets users cue up an instructor-led workout on demand (see Figure 8.21). This means users can have a virtual fitness professional talk them through, for example, a strength training workout at the local YMCA. Other products take the concept even further. Peloton subscribers can tune in to live indoor cycling classes led by professional instructors and compare their performance to other at-home exercisers.

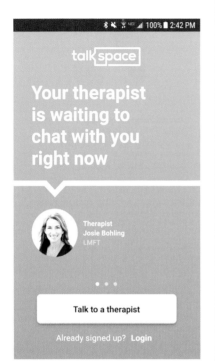

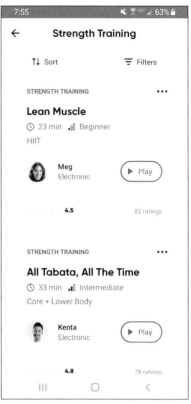

FIGURE 8.20

Talkspace lets people have virtual visits with mental health professionals, which can help combat access issues and feelings of stigma.

FIGURE 8.21

Aaptiv lets users select a prerecorded instructor-led workout to follow at the location of their choice.

When does it make sense to include live experts in digital behavior change products? They can offer the most help to people who are working with complex behavior change problems; patients with MS are a good example. Managing the symptoms of MS may mean changing multiple different behaviors while also coping with the emotional realities of a chronic progressive illness. Likewise, virtual medical consultations are relatively economical compared to visiting a physician in person, so the cost differential makes them appealing to buyers (not to mention the privacy aspect that may be especially appealing if people feel embarrassed about the help they need). And programs like Peloton are geared toward high-end consumers who are willing to pay a premium for a luxury workout experience. For most behavior change designers and most projects, looping in a live expert will be well outside of scope. But when users really need or want that high-touch support, it can provide enormous benefits.

The Upshot: Everybody Needs Somebody

Social support helps support people's feelings of relatedness, which is a basic psychological need. Aside from need support being a generally good thing for people's motivation, social support can provide all kinds of other benefits for your users. Users can turn to their "behavior change buddies" for instrumental help, moral support, accountability, and to refill their existential batteries when times get tough. Working on behavior change with a partner can also deepen people's sense of purpose and commitment.

As a behavior change designer, you can make it easier for people to get social support from their behavior change buddies. Coach them on effective techniques for asking for support and provide functionality in your product to connect people to the right partners. Give people the flexibility to be social on their own terms, even if it means they lurk rather than participate. Offer social networking particularly as the means to the end of strengthening high-quality relationships, rather than for its own sake. And when the situation calls for it, bring in the big guns and let your users work directly with expert coaches, care providers, and other professionals.

Josh Clark almost didn't become a runner. After a series of miserable experiences with running, he hated it and quit. That's not surprising, unless you also know that Josh is the original creator of the Couch to 5k (C25K) running program. Josh didn't stay away from running, and by 1996, he had successfully crossed a few finish lines. With the zeal of the converted, Josh wanted to help other new runners avoid a repeat of his own frustrating experience. C25K is the product of his beliefs that there is a way to learn to run without suffering. Although an obvious C25K strong point is how it incorporates goal-setting and feedback science, Josh said it was actually the rise of social media that propelled it to phenom status. I talked with Josh about the role of social support in C25K's success.

How did support become part of the C25K fabric?

The whole project started as my own personal love letter to running— and in a way to people like me, skeptical would-be runners who had never had particular success, and certainly no particular pleasure in running. I wanted to share how I had actually turned the corner from that. So there was a sense of speaking to an invisible community. I knew that there were plenty of people like me, people who were runners but just didn't know it yet.

I started a small online community website in 1996 and published the original schedule there. But it wasn't until it moved out of that almost private space, and into places like Facebook, that the program itself started to take off. It took on a real grassroots momentum, as people started to create their own conversations around this program. The relative simplicity of the schedule, and the way that it can be easily encapsulated into this week-at-a-time shorthand, made it very easy to make Facebook-size comments. Suddenly, Couch to 5k was everywhere.

What types of support do users need?

I'm reluctant to call these kinds of communities *support groups*, but I think that they have some similarity to them. There's a sense of, "I'm going through something that I am not certain about or feel self-conscious about. And that I would like to find the support of others who have done it." The more that people did the program and had real meaningful experiences with it, the more people sought it out. Those newcomers were willing to be vulnerable and share their experience because they saw that they could get support on the other side of it.

Also, when you're following a running schedule along with other people online, that's different from simply participating in a general-interest discussion. You're always there with your cohort of people, folks who are taking on the exact same challenge that you're tackling in that moment. So, it doesn't have people who are lingering and tired of the conversation, or tired of the new questions. Now, the cool thing is that some people *do* linger because they care about bringing up the next cohort, a pay-it-forward kind of experience. It creates these communities that are just really kind, supportive, and nonjudgmental. For something like finding your way into fitness, when you have experienced only defeats in the past, when you don't have confidence in your own body or ability, or discipline to do it, having a community like that is really powerful.

How can designers support failure?

Sometimes the kindest thing that I see in these Couch to 5k communities is this really warm permission simply to not finish. Or to pick it up later. Or to do that week again. The spirit of Couch to 5k is that it's supposed to be easy and gentle; it suggests a ramp-up that should be sustainable for you. The point is not to follow the program precisely, but to create an experience of growth that is both attainable and sustainable. Designers should always think about providing confidence through early victories— but also in early failures. How might we build kindness and flexibility into our systems to let people find their own level and grow from there?

Josh Clark is the founder and principal of Big Medium, a design studio focusing on forward-looking experiences for artificial intelligence, connected devices, and responsive websites, and the author of several books, including Designing for Touch. *You may have seen him speak at SXSW, An Event Apart, or another of the many stages he graces each year. Although Josh's career has moved beyond his early work with C25K, the program remains a phenomenon.*

Mr. Roboto

Connecting with Technology

P eople don't always need another human being to experience a sense of connection. The deep emotional bonds many people have with their pets proves this. (So might the popularity of the Pet Rock in the 1970s but that's just speculation.) Even Link in *The Legend of Zelda* had an inanimate companion: his trusty sword (see Figure 9.1).

FIGURE 9.1
Even the company of a wooden sword is better than venturing into Hyrule alone.

It's also possible for people to feel that sense of connection in the context of behavior change without having direct relationships with others. By building your product in a way that mimics some of the characteristics of a person-to-person relationship, you can make it possible for your users to feel connected to it. It is possible to coax your users to fall at least a little bit in love with your products; if you don't believe me, try to get an iPhone user to switch operating systems.

It's not just about really liking a product (although you definitely want users to really like your product). With the right design elements, your users might embark on a meaningful bond with your technology, where they feel engaged in an ongoing, two-way relationship with an entity that understands something important about them, yet is recognizably nonhuman. This is a true emotional attachment that supplies at least some of the benefits of a human-to-human relationship. This type of connection can help your users engage more deeply and for a longer period of time with your product. And that should ultimately help them get closer to their behavior change goals.

Amp Up the Anthropomorphization

People can forge relationships with nonhumans easily because of a process called *anthropomorphization*. To anthropomorphize something means to impose human characteristics on it. It's what happens when you see a face in the array of shapes on the right side in Figure 9.2, or when you carry on an extended conversation with your cat.[1]

FIGURE 9.2

The brain is built to seek and recognize human characteristics whenever a pattern suggests they might be there. That means people interpret the array of shapes on the right as face-like, but not the one on the left.

People will find the human qualities in shapes that slightly resemble a face, but you can help speed that process along by deliberately imbuing your product with physical or personality features that resemble people. Voice assistants like Siri, Cortana, and Alexa, for example, are easily perceived as humanlike by users, thanks to their ability to carry on a conversation much like a (somewhat single-minded) person.

Granted, almost nobody would mistake Alexa for a real person, but her human characteristics are pretty convincing. Some research suggests that children who grow up around these voice assistants may be less polite when asking for help, because they hear adults make demands of their devices without saying please or thank you. If you're asking Siri for the weather report and there are little ones in earshot, consider adding the *other* magic words to your request.

---

1 I can write this without embarrassment because research shows I'm not even close to being the only one.

So, if you want people to anthropomorphize your product, give it some human characteristics. Think names, avatars, a voice, or even something like a catchphrase. These details will put your users' natural anthropomorphization tendencies into hyperdrive.

Everything Is Personal

One thing humans do well is personalization. You don't treat your parent the same way you treat your spouse the same way you treat your boss. Each interaction is different based on the identity of the person you're interacting with and the history you have with them. Technology can offer that same kind of individualized experience as another way to mimic people, with lots of other benefits.

Personalization is the Swiss Army knife of the behavior change design toolkit. It can help you craft appropriate goals and milestones, deliver the right feedback at the right time, and offer users meaningful choices in context. It can also help forge an emotional connection between users and technology when it's applied in a way that helps users feel seen and understood.

Some apps have lovely interfaces that let users select colors or background images or button placements for a "personalized" experience. While these types of features are nice, they don't scratch the itch of belonging that true personalization does. When personalization works, it's because it reflects something essential about the user back to them. That doesn't mean it has to be incredibly deep, but it does need to be somewhat more meaningful than whether the user has a pink or green background on their home screen.

Personalized Preferences

During onboarding or early in your users' product experience, allow them to personalize preferences that will shape their experiences in meaningful ways (*not* just color schemes and dashboard configurations). For example, Fitbit asks people their preferred names, and then greets them periodically using their selection. Similarly, Lose It! asks users during setup if they enjoy using data and technology as part of their weight loss process (Figure 9.3). Users who say yes are given an opportunity to integrate trackers and other devices with the app; users who say no are funneled to a manual entry experience. The user experience changes to honor something individual about the user.

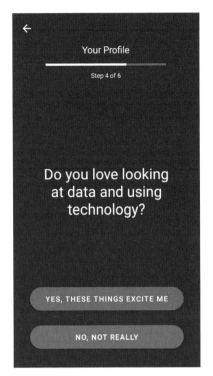

FIGURE 9.3

Lose It! gives users an opportunity to share their technology preferences during onboarding and then uses that choice to shape their future experience.

If you can, recall back to ancient times when Facebook introduced an algorithmic sort of posts in the newsfeed. Facebook users tend to be upset anytime there's a dramatic change to the interface, but their frustration with this one has persisted, for one core reason: Facebook to this day reverts to its own sorting algorithm as a default, even if a user has selected to organize content by date instead. This repeated insistence on their preference over users' makes it less likely that users will feel "seen" by Facebook.[2]

---

2 The difficult-to-escape Facebook algorithm sort also generates legitimate complaints about repetitive content, not having time-sensitive posts displayed at the right time, and intensifying information bubbles.

Personalized Recommendations

If you've ever shopped online, you've probably received personalized recommendations. Amazon is the quintessential example of a recommendation engine. Other commonly encountered personalized recommendations include Facebook's "People You May Know" and Netflix's "Top Picks for [Your Name Here]." These tools use algorithms that suggest new items based on data about what people have done in the past.

Recommendation engines can follow two basic models of personalization. The first one is based on products or items. Each item is tagged with certain attributes. For example, if you were building a workout recommendation engine, you might tag the item of "bicep curls" with "arm exercise," "upper arm," and "uses weights." An algorithm might then select "triceps pulldowns" as a similar item to recommend, since it matches on those attributes. This type of recommendation algorithm says, "If you liked this item, you will like this similar item."

The second personalization model is based on people. People who have attributes in common are identified by a similarity index. These similarity indices can include tens or hundreds of variables to precisely match people to others who are like them in key ways. Then the algorithm makes recommendations based on items that look-alike users have chosen. This recommendation algorithm says, "People like you liked these items."

In reality, many of the more sophisticated recommendation engines (like Amazon's) blend the two types of algorithms in a hybrid approach. And they're effective. McKinsey estimates that 35% of what Amazon sells and 75% of what Netflix users watch are recommended by these engines.

> **TIP** DON'T OVERWHELM
>
> Remember from Chapter 3, "It's My Life," that giving people too many options—something Netflix is notorious for—can overwhelm them. If you're incorporating any sort of algorithmic recommendations in your design, consider capping the number of options at any given time so that your users don't suffer "analysis paralysis." And if you can, limit the number of related options so that users aren't trying to compare the merits of sixteen chicken soup recipes.

Sometimes what appear to be personalized recommendations can come from a much simpler sort of algorithm that doesn't take an individual user's preferences into account at all. These algorithms might just surface the suggestions that are most popular among *all* users, which isn't always a terrible strategy. Some things are popular for a reason. Or recommendations could be made in a set order that doesn't depend on user characteristics at all. This appears to be the case with the Fabulous behavior change app that offers users a series of challenges like "drink water," "eat a healthy breakfast," and "get morning exercise," regardless of whether these behaviors are already part of their routine or not.

When recommendation algorithms work well, they can help people on the receiving end feel like their preferences and needs are understood. When I browse the playlists Spotify creates for me, I see several aspects of myself reflected. There's a playlist with my favorite 90s alt-rock, one with current artists I like, and a third with some of my favorite 80s music (Figure 9.4). Amazon has a similar ability to successfully extrapolate what a person might like from their browsing and purchasing history. I was always amazed that even though I didn't buy any of my kitchen utensils from Amazon, they somehow figured out that I have the red KitchenAid line.

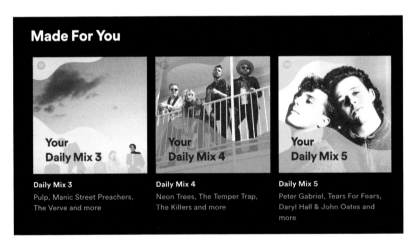

FIGURE 9.4
Spotify picks up on the details of users' musical selections to construct playlists that reflect multiple aspects of their tastes.

A risk to this approach is that recommendations might become redundant as the database of items grows. Retail products are an easy example. For many items, once people have bought one, they likely don't need another, but algorithms aren't always smart enough to stop recommending similar purchases (see Figure 9.5). The same sort of repetition can happen with behavior change programs. There are only so many different ways to set reminders, for example, so at some point it's a good idea to stop bombarding a user with suggestions on the topic.

Don't Be Afraid to Learn

Data-driven personalization comes with another set of risks. The more you know about users, the more they expect you to provide relevant and accurate suggestions. Even the smartest technology will get things wrong sometimes. Give your users opportunities to point out if your product is off-base, and adjust accordingly. Not only will this improve your accuracy over time, but it will also reinforce your users' feelings of being cared for.

Alfred was a recommendation app developed by Clever Sense to help people find new restaurants based on their own preferences, as well as input from their social networks. One of Alfred's mechanisms for gathering data was to ask users to confirm which restaurants they liked from a list of possibilities (see Figure 9.6). Explicitly including training in the experience helped Alfred make better and better recommendations while also giving users the opportunity to chalk errors up to a need for more training.[3]

---

3 Alfred was so successful that Clever Sense, its creator, was acquired by Google, who quickly sunsetted the app. RIP, Alfred.

Having a mechanism for users to exclude some of their data from an algorithm can also be helpful. Amazon allows users to indicate which items in their purchase history should be ignored when making recommendations—a feature that comes in handy if you buy gifts for loved ones whose tastes are very different from yours.

On the flip side, deliberately throwing users a curve ball is a great way to learn more about their tastes and preferences. Over time, algorithms are likely to become more consistent as they get better at pattern matching. Adding the occasional mold-breaking suggestion can prevent boredom and better account for users' quirks. Just because someone loves meditative yoga doesn't mean they don't also like going mountain biking once in a while, but most recommendation engines won't learn that because they'll be too busy recommending yoga

FIGURE 9.6

Alfred included a learning mode where users would indicate places they already enjoyed eating. That data helped improve Alfred's subsequent recommendations.

videos and mindfulness exercises. Every now and then add something into the mix that users won't expect. They'll either reject it or give it a whirl; either way, your recommendation engine gets smarter.

TIP SAY YOU'RE SORRY

With due respect to Ali MacGraw, love means having to say you're sorry if you've screwed up and want to keep your users happy. Mistakes are inevitable. Most people realize that. Owning up and saying something like "Oops! Our algorithms made an error" will help users move past any negative emotions they felt at being misjudged.

Personalized Coaching

At some point, recommendations in the context of behavior change may become something more robust: an actual personalized plan of action. When recommendations grow out of the "you might also like" phase into "here's a series of steps that should work for you," they become a little more complicated. Once a group of personalized recommendations have some sort of cohesiveness to systematically guide a person toward a goal, it becomes *coaching*.

More deeply personalized coaching leads to more effective behavior change. One study by Dr. Vic Strecher, whom you met in Chapter 3, showed that the more a smoking cessation coaching plan was personalized, the more likely people were to successfully quit smoking. A follow-up study by Dr. Strecher's team used fMRI technology to discover that when people read personalized information, it activates areas of their brain associated with the self (see Figure 9.7). That is, people perceive personalized information as self-relevant on a neurological level.

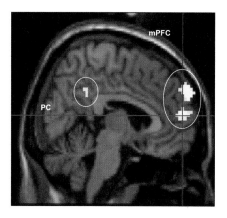 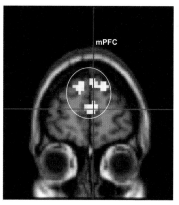

FIGURE 9.7
This is an fMRI image showing activation in a person's medial prefrontal cortex (mPFC), an area of the brain associated with the self. The brain activity was recorded after showing people personalized health information.

This is important because people are more likely to remember and act on relevant information. If you want people to *do* something, personalize the experience that shows them how.

From a practical perspective, personalized coaching also helps overcome a common barrier: People do not want to spend a lot of time reading content. If your program can provide only the most relevant items while leaving the generic stuff on the cutting room floor, you'll offer more concise content that people may actually read.

Speak My Language

One of the most powerful tools in your arsenal to nurture a sense of relatedness with your users is the content that's part of your product. Using plain language is a best practice. The average American reads comfortably at a grade school level, so crafting content that a fourth grader could understand is a good strategy. More sophisticated readers are rarely offended by simple prose done well, so you can capture a wider range of people with less complex content. The words you choose in your content can help forge a connection with users—or alienate them.

Jargon, for example, can be distancing for your average user.[4] Yet, it's incredibly common in behavior change products. Health behavior change products bandy around words like "hypertension," and financial behavior change products reference "compound interest." In the example from Happify in Figure 9.8, users are told activities will "strengthen neural pathways" and help improve their "ability to be mindful." Does the average user really understand these terms? (Spoiler: Probably not.)

Avoiding jargon altogether would be ideal. However, there are reasons why that may not be possible. For example, it can be important to use accurate and precise terminology for legal or compliance reasons, or because of brand identity guidelines that your company has established. It might also be important for people to learn the right terminology as part of their change process. If that's the case, then define terms on first use in simple language. For example, you might write "hypertension or high blood pressure."

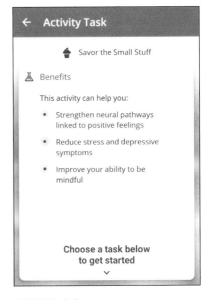

FIGURE 9.8

Terms that make perfect sense to experts, like "neural pathways," "depressive symptoms," and "mindful," are not necessarily meaningful to average users.

---

4 The word "jargon" may be jargon itself. It refers to any terminology that's specific to a profession or group and may not be easily understood by outsiders.

Try to put jargon into a context that helps people understand its meaning.[5] For example, you could say "You can earn more compound interest the earlier you start saving money, so a small amount of savings in your 20s could mean you have more money in your 60s." The example in Figure 9.9 from Clue, a period tracking app, shows another way to add context: The answer options are examples of the term "hormonal birth control" from the question.

How do you know when a term is jargon? Chances are, you have more knowledge than the average bear when it comes to your product's target behaviors. That means you're not the best judge of what's jargon and what's plain language.

Rather than making assumptions, spend some time researching what words people

FIGURE 9.9

If Clue users don't understand the term "hormonal birth control," the answer options will help them figure it out.

actually use to talk about the topics that are part of your intervention. Or involve users directly! When Hopelab created Vivibot for teen cancer survivors, they took nine teen cancer survivors on a weekend retreat and had them help create content that worked for that audience. Other techniques that can help with finding the right terminology are open card sorts and tree testing, or tools like the Hemingway App.

If you can, reflect back the language your users use in the product. If someone calls diabetes "the sugar," it's okay to use that phrase back

---

5 A nonbehavior change example is using the pricing and order of options on a Starbucks menu to know that tall, grande, and venti correspond to small, medium, and large, respectively.

to them. At the very least, you can acknowledge that you understand they may use different words than you do to describe the concepts in your product.

Using the wrong words can cause your users to disengage early in the process, before you really have a chance to create a relationship with them. That's why it's so critical to get the words right.

Watch Your Tone

It's not just the words you choose that matters. The voice and tone of your product do, too. Voice and tone are *how* you talk, and they can help your users feel cared for—or neglected.

Voice communicates your product's overall personality, while *tone* is specific to an individual piece of content. Perhaps your product's voice is "a supportive cheerleader, there to offer help and encouragement with a touch of humor." When helping a user pick a goal, the humor might be dialed up. When providing feedback after someone has failed at an attempt to do something, the humor might be dialed back while the supportive element takes the front seat. The product will have a different tone in these two instances that are both true to the overall voice.

Getting the voice and tone of your product right is vital for it to be a technology that people can feel connected to. Content strategists can conduct research to help develop a voice and tone that bridge user needs and brand priorities (including any jargon or terminology likely to crop up). Dr. Alison Darcy of Woebot offers some other suggestions for capturing the right voice at the end of this chapter.

Talk Like a Psychologist

In Chapter 3, you learned about motivational interviewing (MI) and how to use it to get people thinking about what really matters to them. MI is also an excellent tool to help people feel like they're really connecting with your product. It's part science, part Jedi mind trick. Give people room to express their feelings while you reflect back what you heard—no judgment. As people hear their words echoed back to them, they feel heard and understood.

"No judgment" is critical. If people feel like they're being judged, their natural instinct is to defend themselves. That means they're focused on justifying their behaviors instead of thinking about better ways to tackle behavior change obstacles. There's a clinical concept called "unconditional positive regard"[6] that you can bring into the way you create content for your product. If your content reflects unconditional positive regard, it may include requests for clarification or words of affirmation, but not criticisms (at least not of the person). There is an assumption buried in this approach that people want to do the right thing, so it's possible to point out that a behavior or strategy isn't working without necessarily saying anything negative about the people doing them.

One of the earliest digital behavior change interventions used this style of content to help people with symptoms of depression. ELIZA was created in 1966 by Joseph Weizenbaum, an MIT professor. The algorithm generates content that mostly poses questions based on the input provided by the user, as shown in Figure 9.10. In his research paper about the program, Weizenbaum pointed out that "a large part of whatever elegance may be credited to ELIZA lies in the fact that ELIZA maintains the illusion of understanding." That illusion of understanding is what fulfills people's need for connection.

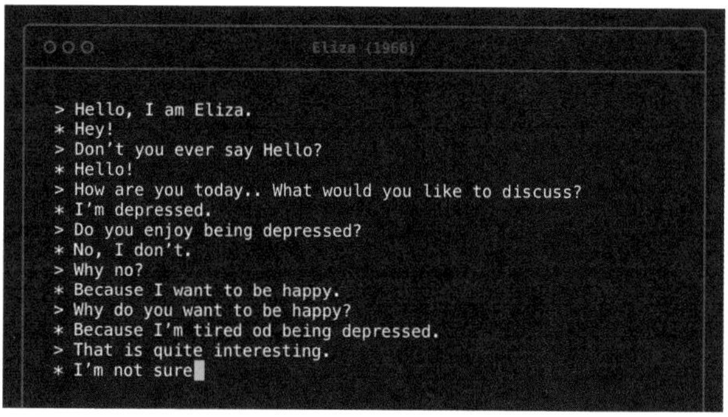

FIGURE 9.10
ELIZA mostly restates content entered by the user to elicit further information. This tactic can help people feel like they're having a meaningful conversation.

```
○ ○ ○                    Eliza (1966)

> Hello, I am Eliza.
* Hey!
> Don't you ever say Hello?
* Hello!
> How are you today.. What would you like to discuss?
* I'm depressed.
> Do you enjoy being depressed?
* No, I don't.
> Why no?
* Because I want to be happy.
> Why do you want to be happy?
* Because I'm tired od being depressed.
> That is quite interesting.
* I'm not sure█
```

---

6 Unconditional positive regard was a core part of psychologist Carl Rogers's approach. If someone is a Rogerian or humanistic therapist, you can expect their style will be nonjudgmental and affirming.

The Turing test evaluates whether a machine can behave in a way that's indistinguishable from a human. ELIZA was one of the first programs to do well on an early version of the Turing test (although people were eventually able to figure out that it was not human). As artificial intelligence has improved, there are still very few programs which pass the Turing test, and even those did not fool a majority of the judges. The takeaway lesson is that bots and AI can serve a valuable function without being mistaken for human beings.

How might this look in a behavior change product? To express unconditional positive regard, your content might borrow ELIZA's strategy of reiterating words and phrases from your users to pose questions. It means that the tone of your content should be warm and gentle, unless there's a clear mandate otherwise—such as a user who's told you they respond well to a "drill sergeant" style of coaching. The example from Nagbot in Figure 9.11 shows how you can ask users if they prefer a more or less aggressive style of feedback. And it means that corrective feedback should be couched in an assumption that people can and want to do better.

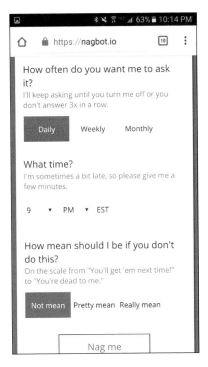

FIGURE 9.11
Nagbot asks users during onboarding how "mean" they'd like the app's nagging to be. The tone of the content adjusts to their preference.

Robots with Personality

Chatbots are perhaps the ultimate manifestation of personality in a digital intervention. Chatbots are automated conversational agents who can interact in real time by responding to user inputs. Some chatbots produce content using "if-then" types of algorithms; increasingly, machine learning is part of their programming so that they can respond to a broader range of topics and learn from previous conversations.

Chatbots pop up all the time (literally) in online customer service scenarios. On many websites, a window will open with an offer to help with any questions—these are chatbots. Chatbots can supplement or replace the need to use a stand-alone app. For example, Uber now lets people order rides by chatting with its Facebook Messenger bot, no app needed, while Spotify's bot lets you send song suggestions to friends to play in the app. Some chatbots are stand-alone programs, which is the case for many of the chatbots being used in behavior change.

Chatbots are good behavior change tools because they can coax a lot of rich information from users. Research shows that people are surprisingly comfortable sharing personal information with chatbots and other digital tools. They may provide more information than they would to a doctor, for example. Studies that have compared what people have told digital interventions with verifiable data like medical records have also found that people are pretty honest when they interact with a machine. It makes sense: computers can't judge the way people judge, so it feels a little easier to share embarrassing information or own up to "bad" behavior. And people who have taken the time to use a behavior change program often want to achieve results, so it's worth their while to provide the best data they can. All of this means that chatbots and conversational agents have promise as a way to engage people with digital behavior change.

For health behavior change, chatbots can help people with some problems that they might not otherwise receive care for. Mental health, including anxiety and depression, is one example. Many people feel embarrassed to admit having mental health issues, which makes it harder for them to seek care. When people do try to find help, for example by making an appointment for talk therapy, they may struggle to find someone or to pay for ongoing care. In many parts of the United States, there aren't enough professionals providing help for all the people who need it. So having some digital tools that can help alleviate anxiety or symptoms of depression is important.

Two chatbots that help people cope with mental health issues are Vivibot and Woebot. Vivibot was created by Hopelab to help teens who have survived cancer cope with their emotions. Many of these teens don't know other young survivors, so finding someone who understands what they're going through is a challenge. Enter Vivi. Vivibot monitors users' mental well-being and offers helpful tips and activities, while also offering people a space to vent.[7] Vivibot's personality is made evident by the emoji peppered throughout its text (see Figure 9.12).

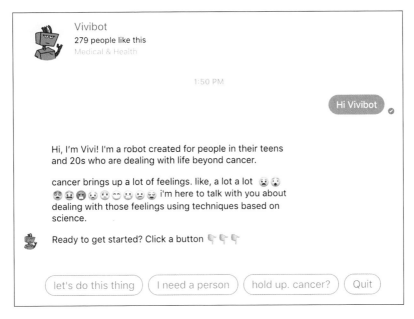

FIGURE 9.12
Vivibot offers teen cancer survivors a place to share their feelings and receive coaching on coping with tough emotions.

Woebot is targeted toward adult users struggling with anxiety and mood issues. It incorporates a scientific approach used by many therapists called "cognitive behavioral therapy" or CBT. CBT teaches people that their thoughts are connected to their feelings, so questioning negative thoughts like "I'm worthless" can help improve feelings. Woebot actually uses the fact that it's a conversational agent to introduce the core concepts of CBT and how the things people say and think can affect their mental health (see Figure 9.13).

---

7 Research suggests that teen cancer survivors actually enjoy chatting with Vivibot and show small reductions in anxiety as a result of doing so.

It's worth noting that both Vivibot and Woebot, along with many other chatbots, offer users a mix of multiple-choice responses and free text. The multiple-choice responses help constrain the flow of discussion within topic areas the chatbot is able to handle, while having at least some free text options helps it feel more like a real conversation. As machine learning and natural language processing capabilities improve, the need for multiple-choice responses will diminish. Even now, chatbots do a decent job with natural language processing. Woebot didn't know the implied meaning of "Netflix and chill," but it figured out it had something to do with relaxing (Figure 9.14).

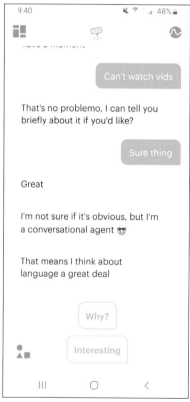

FIGURE 9.13
Woebot uses its status as a conversational agent to segue into a discussion of how language can affect mental health.

FIGURE 9.14
Woebot was able to figure out that "Netflix and chill" had something to do with relaxing and respond appropriately.

If your product's success hinges on people sharing sensitive information, digital may be a good channel to collect it. Just remember that the more personal information you request, the greater your responsibility to treat that data responsibly. In fact, users' willingness to disclose is directly proportional to their belief that their data is private. In the next chapter, you'll learn how to build users' trust so they feel comfortable and confident sharing data.

Two-Way Transmission

You may decide you'd like to have a social media presence for your product. The account identity might be your company (e.g., Delta Airlines), your brand (e.g., Dove), or your specific product (e.g., Gmail). Using social media can be an effective strategy to cultivate a relationship with users outside of your product experience—but it needs to be done well in order to work.

A major error that companies make frequently is to use social media as a marketing channel only. Tools like Twitter, Facebook, and Instagram have social interaction inherent in their design. When companies create a brand presence on those platforms but then don't use any of the conversational features, they miss an opportunity to build a more meaningful relationship with their users (and may miss out on converting nonusers as well).

It's important that your online presence exists in more than just a sales capacity. Some of the other ways that products, brands, or companies can use social media include:

- Sharing stories from users or employees ("Meet our lead engineer.")

- Providing company history or information ("We were founded three years ago today!")

- Revealing parts of the product development process ("See how our team chose the name for our new app.")

- Linking to product-relevant articles ("Learn more about what happens to plastic recycling.")

And, of course, social media can be used to interact directly with people.

By interacting with users (and potential users) in a more meaningful way, you can trigger that anthropomorphization that lets humans feel related to nonhuman entities. You're establishing a personality for users to attach to. Whether you've established a company or a product presence on social network platforms, you can connect more effectively with users through both *reactive* and *proactive* communication.

Reactive communication comes from responding to users' posts that mention either you or a related topic. Sometimes this takes the form of reactions to user complaints or suggestions. Although this may feel demoralizing to designers, it's an opportunity to respond to user disappointment in a positive way that builds loyalty. Take the example from WW (formerly Weight Watchers) in Figure 9.15, where they responded to people confused by a new rewards program for hitting weight loss milestones.

FIGURE 9.15
WW replied to cus-
tomer tweets to clarify
confusion over a new
rewards program.

Proactive communication happens when people search for mentions of their company, or product, or key terms related to their focal area and start a conversation with a relevant reply. These conversations can help people feel connected to a product, brand, or company. A nice bonus to proactive interactions on social media is that they're less likely to involve complaints or customer dissatisfaction. Rather, they offer a chance to show how your product could help fit into something a person is already doing.

Authenticity, Again

Authenticity is important to help people feel a sense of purpose and fulfillment, and to cultivate genuine relationships between people. It's not surprising that authenticity in a product can also help people latch onto it. Don't overly sanitize your product. If a little humor or weirdness bubbles up, roll with it (within reason). People like feeling they've gotten a peek at something real. Check out the customer reaction to Robinhood's quirky privacy notice in Figure 9.16—and how the Robinhood social media manager owns it.

FIGURE 9.16 Robinhood shows an authentic personality, and this customer loves it.

The Upshot: People + Product = Love

You don't need to incorporate actual interpersonal social support in your product to nurture people's feelings of connectedness. A well-designed product lets people form emotional attachments to it that can offer at least some of the benefits of a person-to-person relationship. The keys are to make it easy for people to anthropomorphize, or think of the product in humanlike ways, and to design an experience where users feel personally seen and understood. To the extent that people feel like a product works for their unique needs, they'll feel an attachment to it.

Alison Darcy didn't set out to change behavior. She studied psychology as an undergraduate, but then worked as a software engineer. When the dot com boom in the early 2000s shook the industry, Alison began talking to a friend in the nonprofit world about what an online support group for people with eating disorders in rural areas might look like. That experience led Alison to turn away from postgraduate work in cognitive neuroscience and earn a Ph.D. in treatment design and development instead.

She pursued her interest in digital interventions that are fun to use yet helpful for mental health, including the development of a game called Zombies Got Issues. *(It turns out players were more interested in killing the zombies than in helping them correct distorted thought patterns.) Eventually, Alison hit on the idea of using a conversational agent to help people track mood, and Woebot was born. Alison told me about how she blends clinical science and engaging design, and her tips for other designers to create technology people can love.*

How can you craft a safe conversational space?

We found that when we asked young women, "Is there anything that you would never share with another person?" only about 40% of them would say yes. They would say things like, "No, I tell my best friend everything." All the males would say, "Yeah, of course." They would not even understand why you were asking the question. It was so self-evident that they would not share stuff with another person. That's a key insight that led us to Woebot. People always say things like, "Well, if you're feeling upset you should talk to someone." That actually alienates a very large amount of the population for whom that is not feasible, for reasons like emotional barriers, stigma, or self-stigma, or just logistics, for example, if it's 3:00 a.m.

We have 10 design principles, what we refer to as Woebot's core beliefs. Some of them are therapeutic principles like "sitting with open hands," which is this idea that you'd never push somebody to change. You completely honor their decision to change or not. How that works out in a design is that we don't do persuasive tech. We always issue an invitation, "Do you want my help with this, or do you just want to get it off your chest? That's fine, too."

That's also the promise of Woebot. Because it's not a person, you can strip away all of the impression management. Users don't have to think, "How is this person perceiving me right now if I share this thing?" That is a unique kind of opportunity. You can see in our data that people just go straight into stuff that's very personal.

How do you develop a distinctive product voice?

It's really about being human and imbuing some of that into the writing. And the challenge is doing that in a way that's lean and bouncy—so keeping it short and interactive. It was a lot of coming back to first principles. We finally realized that Woebot should never be sarcastic. He should be a little bit more on the humble side because Woebot is going to get things wrong. That fits with his personality. He says, "I am a robot that's trying to learn from people. I'm really young. I'm only two." And so, there are some faults.

Why make a mental health app so upbeat?

Woebot is a fictional character and users seem to be happy to play along. It works because therapy is also a kind of suspended reality. You're inviting people to imagine that their reality now didn't have to be this way. It's fun to step into this fictional character, "Okay, imagine that you really were a robot." Nobody is mistaking Woebot for being real. And it works really well when somebody is doing real work on themselves, to kindle this kind of lighthearted experimental kind of mindset.

Humor can really help the work to continue in a way that's still honest and respectful of where the person's at. You're not making light of things. You'll notice that when somebody is sharing that they're in a difficult space, there's no humor at that point. We're not trying to make jokes or have gags. We think of it as Woebot not taking himself too seriously. Being accessible means it's a bit tongue-in-cheek because the human condition is kind of absurd anyway.

One of our internal metrics that we look at a lot is the number of bad moods improved. If Woebot's taking you through a therapeutic technique, at the end of it he will always check in and say, "How do you feel after that, better, same, or worse?" We're looking for either better or same because we feel those both have therapeutic value. When we deep dive into why people feel the same, a lot of the time it's because they've unveiled more thoughts that they didn't know were there before. That's therapeutic. You're not actually always going to feel tremendously better if you've done very big work in that session. After about 70% of these conversations, however, people do.

 *Alison Darcy, Ph.D., is founder and CEO of Woebot Labs. Previously, she served as a member of the Faculty in Psychiatry and Behavioral Science at the Stanford School of Medicine.*

A Matter of Trust

Design Users Can Believe In

Designing for trust combines two key considerations. The first is tactical: If you want people to use your products, then you have a responsibility to design them so they can be trusted. It's tempting for me to hammer on about the ethical obligations designers have toward their users with respect to their data. Don't worry, I'll resist the temptation (mostly).

The second part of designing for trust is making it easy for users to see you, your product, and any community associated with your product as good actors. It's about communicating your trustworthiness in a way that can be believed. Doing this is vital for achieving your product's goals, assuming they include actually getting people to do something differently as a result of their experience.

The information in this chapter is roughly organized in the order that someone might encounter it when using a new product, going from onboarding to graduation. It might make sense to mix and match how you apply this information, depending on the details of your particular product; don't get too hung up on matching a tactic to a stage of the user experience. For example, I suggest sharing scientific proof points as part of the initial marketing, but that can be appropriate within the user experience as well. Use your best judgment.

Why Trust Is So Important

When it comes to technology, trust is everything. Users who don't trust in a product or the people behind it will be reluctant users at best. More likely, they'll be nonusers. And woe to you if your users feel betrayed by their experience with your product. It's much harder to regain lost trust than to earn it for the first time.

What does it mean for people to trust a product? It includes confidence that they'll be treated fairly in their experience; their data, including any payment information, will be kept secure; and they won't be hit with hidden charges. It also means that they can count on the product to do what it says it will—that the product's marketing isn't just for advertising purposes, but communicates expectations accurately. If there are live human beings involved in supporting the behavior change, trust extends to them as well—that they work in good faith and are capable of offering the promised support. Finally, trust includes confidence that the product works. In the case of a behavior change product, users who trust a product believe it offers a legitimate protocol that is appropriate for people like them.

The importance of trust multiplies when the technology touches sensitive areas of people's lives. Many behavior change products ask people to put themselves into a vulnerable position, such as when Woebot queries symptoms of depression or Couch to 5k gets newbie runners on the road. Sometimes users are asked to share really personal information (hello, symptoms of depression!). That means reflecting on topics that might be more comfortably ignored. Providing the information that feeds behavior change puts data out into the world that may be embarrassing in the wrong hands, or put people into harmful financial or legal situations. The one-two combo of people entrusting you with their personal data and giving you access to their experience during the difficult process of behavior change means that the trust stakes are especially high.

The process of building and then nurturing trust begins with your very first encounter with users, which usually takes place outside the product itself.

Hello, My Name Is

From the very beginning of the user experience, it's important to establish that your product, the people behind it, and the people supporting and using it can be trusted so that people are willing to engage with it. Otherwise, the path to product goals is broken. As the former U.S. Surgeon General C. Everett Koop said, "Drugs don't work in patients who don't take them." Similarly, your behavior change product won't work if people don't trust it enough to use it.

Before people can use a product, they need to become aware of it. There's much you can do in your marketing and outreach to establish credibility before a user enrolls. People's decision whether or not to give your product a try rests on how solid they believe its foundation is. They'll want to know what benefits they can expect, what the process of being a user is like, and what costs are associated with it.

A lot of this trust-building may take place outside of your product. Your website and other marketing materials are important tools for creating an initial aura of credibility. Successfully introducing your product, its value proposition, and the reasons why it can be trusted is an essential first step in getting people to become users.

The Science Behind

You can add credibility to your product by revealing some of the scientific foundations on which it's built. Many behavior change products have a rigorous rationale behind them. For example, the Woebot mood bot you met in the last chapter uses cognitive behavioral therapy (CBT) to help people cope with symptoms of depression. A retirement planning program uses a financial model to estimate how different savings strategies will help users get their account balances to a target amount by a planned retirement date. The advice in these programs is not randomly chosen, but carefully crafted to bring a proven process to life.

Sharing the foundations of the product, especially if it helps people achieve very meaningful outcomes like financial security or carries a great potential risk if things go awry, can help people trust it enough to become users. If your behavior change intervention deals with health or finances, in particular, consider exposing some of the background rationale to users.

Of course, the average user isn't likely to dig through academic papers in order to verify that your science is solid. Most people will be reassured just knowing that the underlying science exists, and perhaps seeing two or three bullet points describing it. They're likely to be put off if they're forced to endure more detailed explanations. If you want to include details for more sophisticated or inquisitive users, consider following a model like Happify (Figure 10.1), where they put a research summary and a list of citations behind a link. The summary is suitable for a general audience, while geekier users can make use of the reference list to dig up the original research.

Establishing the scientific foundation of your product is also an opportunity to share outcomes if you have them. If your product has been in the market long enough to have real user outcomes, or if you've run your own pilot or research studies, tell people about it so they can feel confident that it might work for them too. Again, you don't need to delve into the methodological nitty-gritty. In Figure 10.2, Shapa advertises how much weight its users lost compared to a nonuser comparison group without getting into much more detail; the headline is compelling on its own.

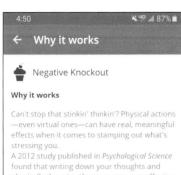

🧁 Negative Knockout

Why it works

Can't stop that stinkin' thinkin'? Physical actions —even virtual ones—can have real, meaningful effects when it comes to stamping out what's stressing you.

A 2012 study published in *Psychological Science* found that writing down your thoughts and physically throwing them away was an effective way for people to banish bothersome thoughts. In the study, students wrote negative or positive thoughts about their body image on a piece of paper. Half of them were instructed to throw away the papers in the trash, while the other half held on to them. Afterwards, when asked to rate their body image, those who kept their written thoughts were affected by what they wrote, whereas those who'd tossed their thoughts away were not. [H-118]

A similar study instructed participants to type their thoughts and save it in a file. Turns out, those who dragged the file into the computer's recycling bin were less affected by those thoughts than people who saved the file to a disk or merely *imagined* moving the file to the recycling bin. [H-118]

When representations of your negative thoughts disappear—even if it's temporary—it's easier not to think about them, explained the researchers. So get your negative thoughts down—and get ready to knock 'em out, one by one!

---

happify™ ☰

Science of Happiness Research

🧁Savor

S-1: Mindfulness strengthens parts of the brain connected with emotion regulation, happiness, learning & memory, and perspective-taking

- Kilpatrick, L.A., Suyenobu, B.Y., Smith, S.R. et al. (2011). Impact of mindfulness-based stress reduction training on intrinsic brain connectivity. NeuroImage.
- Hölzel, B.K., Carmody, J., Vangel, M. et al. (2011). Mindfulness practice leads to increases in regional brain gray matter density. Psychiatry Research.

S-2: Health benefits of savoring

- Weinstein, N. & Ryan, R. (2010). When helping helps: Autonomous motivation for pro-social behavior and its influence on well-being for the helper and recipient.

FIGURE 10.1

Happify makes detailed information about the science behind it available to users, but they must click through to find it.

shapa ☰

Average
weight loss

5.8 lbs

Of the Shapa users who lost weight, they lost an average of 5.8 pounds in 12 weeks.

GET STARTED

FIGURE 10.2

Shapa's website lets people considering signing up how much weight their users lost in a research study compared to non-users. More detail isn't necessary for most audiences.

Trust Us, We're Doctors

Another way to establish credibility is highlighting the foundations of your product, including any institutional affiliations or scientific backstory. Some behavior change products are commercial enterprises associated with research centers or universities. For example, BecomeAnEx was built in partnership with the Mayo Clinic, a world-respected healthcare institution. Including their name on the website, as seen in Figure 10.3, boosts BecomeAnEx's credibility among people searching for a good smoking cessation program. Elsewhere on the site, interested visitors can read about how the Truth Initiative and the Mayo Clinic Nicotine Dependence Center partnered in 2008 to launch the program.

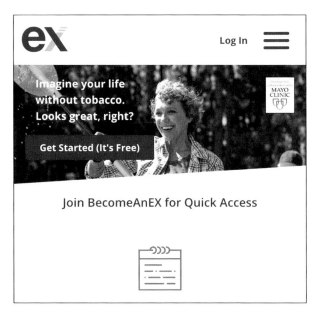

FIGURE 10.3
The Mayo Clinic logo on the BecomeAnEx website is small, but does heavy lifting in terms of establishing the program's credibility.

You can also share the bios or qualifications of the people who worked on building the product or are represented in it (like if there is a coach or expert who sometimes appears in the program itself). Sometimes those institutional affiliations belong to an individual on the team instead of the organization as a whole. If your lead behavior change designer is also a faculty member at a top university, tell people that. The websites for behavior change programs often include a section about their team of experts; the purpose is to increase people's trust in the program.

It's also possible to describe the qualifications of people associated with the product en masse, like Curable does (see Figure 10.4). Instead of providing individual bios for all of the people involved in developing it, the pain management program Curable simply says they include former pain sufferers, physicians, and pain experts. This can be enough to reassure potential users that the program will be appropriate for someone like them.

Show Off Satisfied Customers

Testimonials from other users can help build credibility. Testimonials are evidence that people have used and liked your product. They're also an opportunity to share the benefits and outcomes that people might experience if they use it. For example, DietBet includes before-and-after photos of users along with how much weight they lost and how much money they won within the first few screens of the app (see Figure 10.5).

FIGURE 10.4

Curable describes itself as being built by former pain sufferers, physicians, and pain experts without going into detail about each individual contributor.

The more that testimonials come from people who resemble your target users, the more compelling they'll be. This is especially true if your product is specialized—for example, if it deals with changing behaviors related to a specific health condition or is intended for people who are part of a marginalized community. Reflecting an understanding of these users builds trust, and failing to do so can damage it.

TIP CELEBRITY ENDORSEMENTS

Celebrity endorsements exist in a strange no man's land. They might be user testimonials, as is the case with Oprah Winfrey and Weight Watchers. Or they might be purely business arrangements—does anyone really think Kim Kardashian wears Skechers? There's no doubt that a famous name can enhance the appeal of a product, but tread carefully in the world of behavior change where credibility is king.

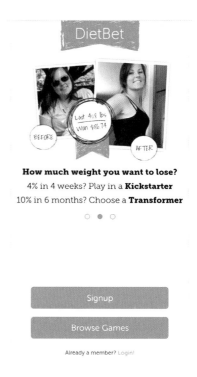

If your product has a B2B sales model, where your customers are not the end users, it's still important to include at least some information on your website to help end users understand what your product does. Many potential end users will do some research before enrolling in a product offered to them by an organization. Having information available for them about how your product works, what their experience will be like, and what benefits they might gain will make them more likely to take the plunge and sign up.

FIGURE 10.5

DietBet includes photos of users who have successfully lost weight and won money as a proof point for people considering joining.

Provide Itemized Receipts

In Chapter 3, "It's My Life," you learned about the importance of letting your users make meaningful choices for themselves as a tool to help them stick with behavior change for the long haul. One of the reasons that user autonomy is associated with long-term engagement is that it helps people trust the products they use. Unfortunately, many behavior change products are marketed in a way that tries to make them sound as appealing as possible without revealing some of the challenges that might be involved. If your users don't already realize behavior change can be hard, they will know it once they start working on it. And if you've concealed that from them, they'll lose faith in you.

It's important to be up front about the nuts and bolts of the behavior change program. Clear expectation setting gives people the information they need to freely choose to become users or not, and ensures that they don't lose trust because their expectations aren't met. What type of information will users be asked to share and why? Will users need special equipment in order to participate, like a fitness tracker or a scale? What fees are associated with the product?

Naturally, it's more difficult to get people to use an app that costs something than one that's free. Even if the price tag is only a few dollars, most people are more reluctant to pay for something than use a free alternative. Some product teams have tried to get around this by making their apps free to download and try, but then asking users to upgrade to a subscription in order to continue. The idea is that the product is so good, people will want to pay for it rather than lose access once they've tried it out. In reality, this sort of "freemium" model can leave people feeling tricked and betrayed. I tested a brain-training app called *Elevate,* which had me take a lengthy assessment and then assigned me different cognitive skills to work on. Only after all that up-front work did the app tell me about its $40-per-year cost. Nowhere on its website or in the onboarding process did it mention that there was a (relatively hefty) subscription fee. As you might guess, the negative feedback on the app store skews heavily toward people surprised by the cost.

All Aboard

So now you have an audience of people who have found your value proposition compelling and decided to become users. It's time to onboard them to your product! The onboarding process offers multiple opportunities to reinforce and build on the initial trust that your new users have in you. Specifically, this is the time to get into the fine print of how the product works and how you've done the groundwork to keep your users safe.

People may approach a new product with trust. But if you don't give them reason to sustain that trust, they'll ultimately choose not to stick with the product. Or they'll find workarounds to protect themselves while still getting the benefit of your product, like the people who create Facebook accounts for their pets. These users get a watered-down experience, and you miss out on building an authentic relationship with them. And while people are willing to forgive other people for mistakes, they are far less willing to give apps or digital products a second chance, so it's vital that you don't screw it up. No pressure.

Prioritize Data Privacy and Security

It's not paranoia if you're right. The users who are most hesitant to trust their data to digital products are in many ways logically correct. Companies can be awful about protecting people's data.

Sometimes, it seems to be a combination of companies letting their guard down just as bad actors look for a way in, such as the 2017 incident in which 143 million Americans had their personal financial data put at risk by Equifax. Sometimes, it's companies either not thinking through the implications of how they handle user data or, if you're feeling extra cynical, being more motivated by their bottom line than their users' well-being. Either one of those scenarios could explain the ongoing wave of scandals with Facebook sharing users' data with third-party companies in ways that not only expose their personal information, but also contribute to widespread misinformation.[1]

And sometimes it's just that companies don't keep ahead of what motivated attackers can do to their technology. There are reports that hackers can hijack connected devices with high-frequency voice commands not detectable by human ears, allowing them to violate people's homes. It's not clear if this possibility got any meaningful time on an agenda at Amazon, Apple, or Google when they were first developing their voice assistants.

If you want your users' trust, deserve it. The technical aspect of that is super important, even if it doesn't usually fall on the to-do list of the behavior change design team. Make sure that someone is doing

---

1 I pick on Facebook a bit in this chapter and elsewhere in the book. I don't think they're the only company making these types of errors, but because almost everyone uses Facebook and their missteps have gotten a ton of press, the examples should be immediately familiar.

all the data security stuff. Take it seriously, act quickly to fix problems, and when in doubt, pretend that you're talking about the data of the people you love most in the world and act accordingly. If you're Facebook and you're about to ink a deal with Cambridge Analytica, review the hell out of how your users' data will be used and where it will be shared, and if you can't resist the paycheck, at least have the integrity to ask your users' explicit permission to participate.[2]

Now tell your users about the steps you've taken to protect their data!

Ask Permission Before Sharing Data

There are all kinds of good reasons why you may want to share your users' data with a third party—or receive user data from a third party. Maybe you want to integrate with their fitness tracker so that you can offer more accurate health coaching. Maybe you want to pull spending patterns from their checking account so that you can offer feedback for saving more money. These are helpful things that may make your product more effective for your users.

So by all means, pursue third-party integrations. Just be sure to ask users' permission—their explicit, specific permission—before proceeding. Detail the exact data you'll be providing or requesting along with a rationale for why. And if users decide not to permit the data sharing, and you don't absolutely need the data for your product to work, don't make it hard for them to use the program: that's coercion.

Take It Slowly

As users set up their accounts, you may need to ask them to authorize functionality that makes your product work better, like granting access to photos on a mobile device or connecting to a device.

Ask for permissions slowly. Hitting users with a catchall authorization request doesn't give them an opportunity to think critically about what they want to share and what benefit it will provide back to them. Ask for each permission individually, and if it's not obvious, explain why the product needs that access. Some permissions may be necessary for the product to work; users should know that, so they can choose accordingly. A combination of a legitimate-sounding reason and reassurances that the data won't be misused will go a long way. Strava demonstrates what this might look like by clearly

---

2 Spoiler: This is not how it actually went down.

describing why it asks for location data and the value the user receives from permitting it (Figure 10.6).

De-Mystify Your Legalese

Getting your users to trust your product isn't just about the fun stuff like developing a snappy avatar or writing witty content. It's also about the boring stuff, like privacy policies (which are legally required worldwide, so you're going to have to include one). In an ideal universe, the privacy policy and other legalese would be written in a way that users can easily understand, especially the bottom line impact on their data and security. Unfortunately, it's easier said than done.

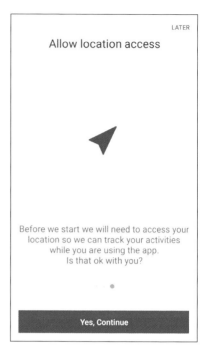

FIGURE 10.6
Strava clearly states the value that users can expect if they allow the app to view their location data.

Why is it hard? Well, it's legalese. The reason product teams include it in their apps isn't because they like it. They're including it because it's legally required. In bigger companies or conservative industries, there's probably an in-house legal team dictating its content. To some extent, the legalese is about CYA more than anything else, which means the kitchen sink approach reigns. Although the audience for the legalese is the end user, this content really isn't *for* the end user. It's lengthy and laden with jargon and technicalities. Oh, and it's often one of the last things done, at the end of a project, when no one has the time or energy to lovingly craft it.

The good news is, if you've done your research and know your market space well, you have a pretty good sense of what fears your particular users might have about using your product. And if you have that information, then your legalese is an opportunity to assuage those fears and put users at ease.

When I built digital health coaching tools, our users had one big fear in common: they were afraid we shared their individual health data with their HR manager or healthcare provider. They worried that using our programs was letting their bosses and doctors see their most private, personal information. And that was stopping some of them from ever using the programs.

In reality, HR managers and doctors never saw any output from our programs that wasn't printed and shared by the user. We'd pull together enterprise-level reports for our customers that showed an aggregate of all their employees or members, but it was impossible to pinpoint any individual's responses. If the health plan or employer only had a few hundred people in their sample, we'd take additional steps to anonymize the data in those reports. For example, we wouldn't give average scores for the accounting department if the accounting department had only three people.

Of course, we explained all of this in our legalese. Specifically, it said that we "may create de-identified, statistical, and aggregate data results for research purposes and for reporting to other organizations, including your sponsor (i.e., your employer, health plan, or healthcare provider). This data reported will not contain personal identifiers and will not identify individual participants." Super clear, right? As you probably imagine, almost no one read this text or really understood it. Like most legalese, ours was a dense block of text that was hard to skim.

We proposed revising the legalese to help new users trust that we weren't going to share their information inappropriately. We pulled that piece about not sharing individual data out of the huge text block and moved it front and center. We rewrote it in simple language. We made some wild design choices involving bold and large font. We repeated our commitment to privacy. "Remember, anything you share is confidential. Any reports we produce use average data from hundreds or even thousands of people. We don't provide any information about you personally to anyone." When we tested the change with users in focus groups, they felt much more confident about using our programs. We helped them trust us.

Did the lawyers love this? Not at first. It was a new way of handling information that in their minds was a vital protection for the company against liability. We had many rounds of review and discussion before our legal reviewers felt comfortable that the simplified privacy statement was as legally protective as the dense, complicated one. One of the concessions was that the original legalese survived, below the revised simplifications. I'm sure there were tens of people who scrolled down to read it.

Pinterest is an example of a product that has an easy-to-read and -understand privacy policy (Figure 10.7). They accomplish this by organizing the content so that it's broken out by topic, written in clear language, and followed by a summary, They make it much easier for their users to trust them. Yes, it's still very text-heavy and not something most people will read in detail, but it's *much* easier to navigate than, say, the iTunes user agreement that has famously exceeded 50 pages at times. Be like Pinterest, not like iTunes.

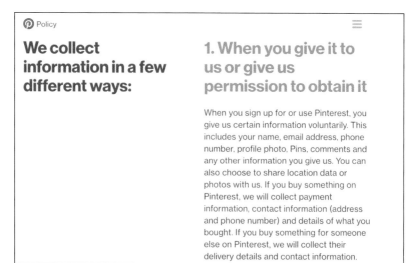

FIGURE 10.7

Pinterest's legalese is relatively easy to read and understand, thanks to how it's formatted and written.

Certificates Galore

Savvy users know to look for certain signs of trustworthiness on websites or apps that collect payment information or other highly sensitive data. It's a good idea to include these in your product. They include things like https instead of http in a website's URL, or the logos of security services like Verisign. Depending on the industry in which you work, it could also include accreditations from organizations like the Better Business Bureau (consumer products), NCQA, JCAHO, or URAC (health care), or J.D. Power or the FDIC (financial services). But proceed with caution: At some point, the lady doth protest too much. One or two of the most critical trust signals is probably enough to give your users confidence.

Deliver on Trust

Now that your users have onboarded to your product, you have an ongoing opportunity to live into the credibility you've established in your relationship with them so far. While the general rule of thumb for trust within a product experience is "Do what you say and say what you do," there are some specific features that can coax user trust. For example, there are trust signifiers you can build into the actual product experience, ranging from the way you respond to user inputs to how easy you make it for users to feel a sense of kinship with your product. If you have a subscription model or ongoing customer support, the way you handle those can also strengthen (or weaken) a user's trust.

Provide Clear and Immediate Value

Earlier, I advocated for being clear with users about why you're asking for a particular permission or piece of data, especially if it is something people might consider private or risky to share. Roobrik provides an excellent model to follow (Figure 10.8). Most of their assessment questions include a sidebar explanation of why that data matters for determining caregiver arrangements. After users have granted your request, it's time to follow through on that explanation.

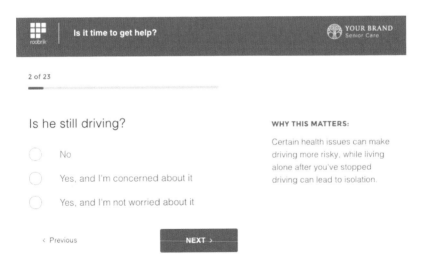

FIGURE 10.8

Each question in the Roobrik assessment is accompanied by an explanation of why the data is important.

As soon as possible, demonstrate back to the user how sharing their information affects their experience within the product. A simple example is from 7 Cups, an app that provides emotional support either via a bot or by connecting users to each other. During onboarding, 7 Cups asked what I like to be called. As soon as I answered, the chatbot began to address me by that name. This is a way to tell me that my data hasn't gone into the void, and reassure me that the answers I provided do add value to my experience in the program.

Depending on what your product does, it's possible to do a more sophisticated version of this as you learn more about your user. TurboTax shows a running total of users' federal and state tax bottom lines that updates with each new piece of information. Other tactics might include sharing normative feedback related to the data ("You're in good company—56% of our users said the same thing.") or an active-listening type response leading into another question ("Wow, that sounds tough. Which of the following best describes how you coped?"). Even the gradual filling of a progress bar can help users feel like their inputs are contributing to a successful product experience.

As a general best practice, have a rationale for every piece of data you ask your users to provide. Ask yourself, "How will collecting this data change the user's experience? How will it affect our understanding of whether and how the product works?" If it's not essential to either craft their experience or inform your measurement strategy, reconsider whether it's appropriate to collect information at all.

Try a Little Tenderness

Behavior change puts people into a vulnerable place. You have an opportunity to react to that vulnerability with sensitivity and respect within the design of your product. Think about a time when you might have shared a secret with someone—something that you felt embarrassed or shy to talk about. If the person you told reacted warmly, that was a very different experience than if they pointed their finger and laughed at you. Similarly, you can help your users feel like they have a safe place to try behavior change tactics, fail, and learn, or you can shame them for not being immediately successful. You can guess how much trust is in each situation.

According to the "broaden and build theory of positive emotions," people's emotional states affect how much they can learn and take risks. If someone is in a negative emotional state, they get narrowly focused on whatever is causing their problem. It consumes all their energy. If they're in a positive emotional state, they can spend their energy widely. They have the capacity to experiment and try new things. They have the ability to trust. This line of thinking suggests that you can help your users actually succeed at the guts of behavior change, which requires some creativity and grit, by supporting their positive emotional states.

Think about how your design can support positive feelings for your users. It may mean injecting moments of levity, like Woebot does (Figure 10.9), with the caveat that any humor shouldn't be mean or make light of serious issues. The content tips from Chapter 8, "Come Together," work well for creating trust as well. Use warm, friendly language and express empathy toward people's situations. Encourage them when their behavior change efforts fall short of their goal, and congratulate them for milestones achieved. Anything you can do to help users feel safe and supported will contribute to their trust in your product.

FIGURE 10.9

The light humor in Woebot can help put people into a positive mood state where they are more willing to extend trust.

Anthropomorphize Like It's Going Out of Style

In the last chapter, you learned that people will *anthropomorphize*, or see an inanimate object as humanlike, given the slightest provocation. So giving your product human characteristics, whether it's an avatar or a name or a tendency to sass, can quickly get users into a relationship state of mind. That's great when it comes to trust, because there's nothing people trust more than their friends.

So it's not surprising that making technology more humanlike often makes it easier for people to trust. In fact, avatars have been put to work in hospitals and health clinics doing jobs that require a lot of trust on the part of patients.

Medical appointments are the best and the worst of times for spurring behavior change. They're the best of times because a lot of people really trust their doctors and are likely to at least listen to, if not follow, their advice. Doctors are also fabulous resources; they're experts, and can help people understand their situation in a much more personalized way than Dr. Google can.

They're the worst of times because doctors have very little time to spend with patients and often aren't able to have a heart-to-heart conversation. Patients may be stressed, scared, or overwhelmed, which makes it hard for them to really listen to what the doctor is saying. Now factor in that doctors are highly trained professionals whose default way of talking about health is crammed with jargon and technical detail. It's not a recipe for understanding. In fact, research suggests that people forget 50% of what the doctor tells them by the time they leave the office.

Enter the avatar.

Timothy Bickmore studies the use of cartoonish avatars (see Figure 10.10) in medical settings. His avatars are animated health professionals who appear in videos that patients can watch on television screens or tablets. The avatars handle some of the in-depth conversations that doctors often don't have time to have. For example, when someone is being discharged to go home from the hospital, they're usually given a laundry list of instructions to follow. In some of Bickmore's studies, patients receive these instructions from the avatar instead.

FIGURE 10.10
The avatars used in hospitals and doctors' offices to give patients discharge instructions are cartoon drawings of healthcare professionals.

The avatar never runs out of time or gets frustrated with patients. Patients can ask it to replay the discharge instructions over and over if they missed something the first time. They can select questions from a menu to get more information. They can toggle the language if they're multilingual and do better hearing it in another tongue.

People like these avatars, and they work. In studies where people talked to avatars during hospital discharge, 74% said they preferred it to getting their instructions from a live person. People who used an avatar to learn about their depression were more likely to feel a "therapeutic alliance" with the avatar than their live therapist—that's jargon for a meaningful sort of provider-patient relationship. Other research has shown that people learn more when they get health information from avatars than when they get it from a person. This is likely both because they can spend more time repeating and studying the information, and because it comes from a trusted source.

These avatars are built to be anthropomorphized. No doubt they also pick up a bit of trust sparkle from being used in a hospital by health professionals which is a *very serious place*. Wherever the trust comes from, people are willing to spend time discussing their intimate health information with avatars, and they feel good about it.

Using multimedia in combination with avatars in your product can help solidify user trust. Research has found that people find audio-, video-, or avatar-based information more believable than text. This even extends to the newest technologies. One study found that people were more likely to trust human-esque navigation systems in self-driving cars over standard models.

Not Too Human

Remember the uncanny valley from Chapter 9, "Mr Roboto"? If technology is too human, it becomes creepy.[3] Going with a cartoon-ish interface like Bickmore's avatars has the advantage of keeping products well away from the edge of the uncanny valley. It can also amp up another effect of technology, which is that people sometimes trust it more than they trust other people.

In the last chapter, I mentioned that research has shown people may be more comfortable sharing personal information with a digital

---

3 Assuming it's not *so* human that it actually fools people, which is creepy for a different reason.

program than with a live human. In fact, telling people that a "virtual coach" is fully automated makes them trust it more than telling them it's piloted by a human. Remember, technology can't give dirty looks when someone confesses to "bad" behavior. And if an app makes a snarky remark about a user, the user can delete it—something that's a lot harder to do to, say, an annoying coach.

Sharing embarrassing information may be easier if there's no fear of (human) judgment. By taking the judgment piece off the table, you can help people feel trust. So perhaps the trick is not so much making your product human as it is making it just human enough.

Let the People Be in Charge

A major reason that people distrust tech tools is they don't feel in control of their experience. Think about your relationships with other people. It's not very satisfying if someone else is always making the decisions about what to do without asking your input or considering your likes and dislikes. While you may trust a person like that in some ways, you probably don't believe they have your best interests top of mind. It's the same with products; when they impose their will on you, you're likely to suspect them of having an ulterior motive.

The design antidote is to offer people opportunities to shape how they use technology. I'll pick on Facebook and its newsfeed algorithm again. The algorithm determines which stories a user sees and in what order. It may prioritize the people that the user interacts with most, but it also may boost certain paid advertisements or stories from advertisers shared by a person's friends. Facebook offers limited control over individual users' newsfeeds—you can sort by recency, but Facebook automatically reverts to its own algorithm the next time you log in, and it's hard to calibrate what information shows up in the feed. One study of Facebook users found they were specifically upset about a perceived loss of privacy when they couldn't override the algorithm. A little bit of control might go a long way here.

Some programs have "expert" algorithms that make recommendations for users. This is common in financial behavior change tools, which can crunch the data about people's bank accounts and goals and offer suggestions on where to invest. The average person is not very good at this sort of calculation, so it's a smart use of technology to get better results. Except research shows that people don't trust these algorithms unless they can peek under the hood. If they're allowed to tinker with the algorithm and make adjustments, their trust skyrockets compared to when they're not allowed to touch it. This is true even though, in most cases, having a nonexpert user make changes to an algorithm like that makes it less effective.

Balancing User Control and Expert Success

People's love for control puts product designers into a quandary because while they do want people to trust their products, they also want them to achieve good outcomes from using them. The adjustments that users make to an expert system could mean less impressive results, which could reduce trust over time because the thing doesn't really work that well. So how can you give people a sense of control while also preventing them from screwing the whole thing up?

Feedback is one tool that comes in handy here. Showing people the impact of their changes on outcomes can help talk them out of bad choices. If they realize that they're harming their own bottom line, they also become less interested in making changes. Basically, if you can prove to people that the expert system will do a better job than they will, they'll lose the urge to tinker with it without losing trust.

Another option is to be very transparent about where recommendations come from without giving people the option to actually adjust the recommendations. Research shows that when people understand how a recommendation algorithm works, they trust it more. Tell users what evidence is considered in making recommendations and why they can believe they're getting good advice. This can also tie back to sharing the scientific foundation of your program, if that's relevant. Giving people explanations of complex processes behind the scenes of your product sets the stage for them to trust in the outcome.

The importance of transparency increases the more your product touches meaningful pieces of users' lives. People are afraid of the black box, especially when the results of whatever happen in there matter a lot. If your product concerns someone's health, money, or home, then it's especially important that they feel like they have some insight into what's happening.

Garbage In, Garbage Out

Incorporating algorithms into your product offers some advantages if you've got a large volume of information and need to find the right pieces to serve to a given user. In behavior change applications, they may recommend action steps toward a goal, for example. Algorithms are more efficient than manual "if-then" programming, and theoretically, they can learn and become more effective as they are used and receive feedback.

Unfortunately, algorithms can go awry if they're developed using problematic data. And so many of the data sets out there are problematic. Data may have been collected in a historical context that either doesn't reflect the current day or the desired future. That's part of what happened when Amazon deployed an algorithm to screen job candidate résumés. The algorithm started penalizing women's résumés, because the historical data set of successful candidates was heavily male. Sometimes data deliberately discriminates against some groups or individuals, and algorithms trained using it will propagate that bias. That may be behind the COMPAS algorithm incorrectly predicting that black defendants are more likely to re-offend than their white counterparts; that particular algorithm was used by judges to determine sentencing, so its error had real consequences for real people.

The issues of data quality in algorithms are too complex to address in detail here, but as a behavior change designer you should be aware they exist. If you are building a product that includes an algorithm, think critically about the training data you use. Try to identify ways in which it might be biased and find ways to correct for that. Monitor the algorithm's output closely, and be ready to intervene if things seem off. The more consequential the output of your algorithm is, the more vigilant you need to be to ensure that your product isn't name-dropped on the evening news.

How May I Help You?

Although it usually lives outside of the actual product, customer service can play a critical role in how much people trust the product. Be available for help. Don't make people search too hard for customer service contact information. If you don't have live help available, commit to a quick turnaround time for a response. Radio silence from customer service doesn't exactly scream trustworthiness.

If you have a subscription model where users need to pay an ongoing fee in order to use your product, you are dealing with another personal topic beyond the behavior change itself. People tend to be sensitive where their money is concerned. If you have the ability to charge someone's credit card on file, you've got an express lane to trust. The good news is, if someone paid the initial subscription fee, they had enough baseline trust to enter their credit card number in the first place. However, any misstep in how you handle the subscription payments could easily shatter that trust. Aside from vigilantly protecting users' financial data, there are a few rules of thumb for maintaining their trust when money is involved.

First, provide warning before any charges are processed. This ensures that no one will be surprised when they open their credit card statement. It also gives users a chance to either cancel, if they are no longer interested in the product, or recommit to it. If they do choose to stay on, the fact that they did it deliberately may strengthen their interest in using the product.

Second, make it easy to cancel. Yes, the best-case scenario is that people don't cancel. But if they do want to leave and you make it hard for them, you're not likely to truly stop them. It's only a delay. And while users are struggling to quit, they're building a reservoir of animus toward you. That's not good for anyone. There are ways to say goodbye with grace and leave users with warmth in their hearts.

How to Say Goodbye

There will likely become a point when your users stop being your users. They may move on because they've achieved their behavior change goal and don't need to work on it anymore. They may lose interest. Or they may choose to use a competitor's product, or perhaps they've gotten access to your product through their employer and now they're changing jobs. It happens.

Losing users hurts, but that doesn't mean you should try desperate measures to keep them. You can design for user exits in a way that leaves the door open for them to reengage in the future. If you end the relationship on a positive note, people are much more willing to recommend your product to others or consider using it again themselves if they need it. A gentle goodbye also takes advantage of two cognitive biases people have: the *recency effect*, where more recent incidents are more easily remembered, and the *Pollyanna principle*, where people remember pleasant events more accurately than unpleasant ones. Leave a good taste in people's mouths.

Making cancellations easy is one way for a graceful exit. Other considerations include what you do with a user's account data when they cancel. Whether you delete it from your database or maintain it in a way where a user can spin their account back up, be clear in your communications so that people understand. Don't try to trick people into staying by using dark patterns to muddy the quit sequence. And throughout, keep the tone of your content consistent with the rest of your product experience and be respectful of the user's choice. The goodbye that Stash provides its investors when they close an account strikes the right note (Figure 10.11). It laments the end of the relationship, but wishes the user well and provides one last bit of advice.

The Upshot: Users Trust an Open Book

At every stage in the user experience, you have opportunities to create trust in your product and your team. From the way that you market and introduce your product, to the onboarding process, to the privacy policy and other legalese, all the way through people actively using the product, the choices you make can create confidence in your product. At the core of trust is transparency: being open and honest about how your product works, what users can expect, and why you ask them to do certain things.

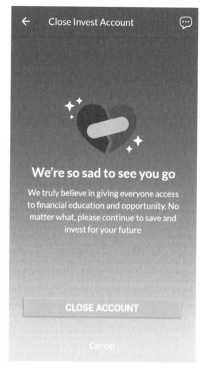

FIGURE 10.11

Stash is sad to see its users quit, but leaves them with a kind message.

Of course, transparency only works when you're able to share genuinely positive information. It's vital to design with respect for your users and their situation, and to carefully protect their data and privacy. By paying attention to potential pitfalls and safeguarding against them, you can keep your users from ever having a reason to lose trust in you. Think of trust as an ongoing conversation between you and your users. As you learn more about your users, their experience should evolve to reflect what you now know about them and reinforce that the information they're giving you is being put to meaningful use.

Sara Wachter-Boettcher seems fearless, at least to me. I became familiar with her after reading the book she coauthored with Eric Meyer, Design for Real Life, *in which she grappled with designing for people on the worst days of their lives. The book, which you should read, challenges designers to anticipate that the people using their products might be in the midst of tragedy, heartache, or pain, and to make choices accordingly. Her next book,* Technically Wrong: Sexist Apps, Biased Algorithms, and Other Threats of Toxic Tech, *is a smart analysis of the ethical follies infecting the tech world. There are a lot of them.*

I see Sara's insights on biases, oversights, and broken processes as being at the very heart of designing for trust. It's when people make the mistakes she cautions against that they lose their users' trust.

How can designers learn to predict pitfalls?

Start by broadening your context: What are you designing? What are you making? Now consider, what is the history of that space? Which systems already exist there? For example, if you are going to collect personal data, then read up on the history of how we have collected personal data in the past—including offline. For example, I don't how many people know that the United States used census data, which we've been told is anonymized, when it rounded up Japanese Americans and put them in internment camps. If people spent more time examining these histories, they might make different choices about what they are and are not comfortable collecting. But we don't have enough of those discussions, because they're easy to deprioritize.

We are really doing our users, the world, and our future selves a disservice by not thinking about worst-case scenarios all the time. It goes against the sunny optimism we see so much of in our field. People want to think about what the positive outcomes might be, and as a result, that muscle of actually thinking about the negatives is not very strong. They're just not used to doing it. But I would say it's just as important of a skill as anything else, and so we should be building it every day.

How do designers end up violating trust?

A bunch of things have kind of rolled together to create a perfect storm right now. We are living through a period of pretty intense venture capital-funded technology. You are not particularly incentivized to understand the implications or long-term consequences of the work that you're doing. The obsession with Agile and "minimum-viable product," with sprints, with a "move fast break things," "get to an alpha launch sooner," ethos—it's all tied in. It really incentivizes short-term decision-making. People are incentivized not to think about consequences. Then, because of those incentive structures, because of cultural norms, even when people want to do it, they're just not that good at it. You're not good at stuff you don't do.

People really get bogged down with the idea that everything they're thinking about has to scale. The obsession with talking about scale has been a disservice in a lot of ways. Sure, the problems we are facing are at a tremendous scale, but things don't work at the large scale unless and until they work at the small scale.

How can you change your work culture to support trust?

You can't change anybody but yourself. Fundamentally, you can educate other people, you can hope other people change, and you can encourage them to change. But the only person you can really control is yourself. I want to encourage people to try to avoid getting so overwhelmed and just focus in on, "What are the choices that I'm making on a day-to-day basis?" If you have choice about where you work, what a privilege! So what are you doing with that choice?

I also really believe in the power of being tough on the things that you love. The more strongly you think, "Oh gosh, this is awesome," the more you need to critique it. I think of it like this: I'm watching out for my future regrets. Is this something we're going to be proud of in 10 years—or even in 6 months? Be honest with yourself—and be honest when you are making a moral compromise because it's easier or better for your career. Because ultimately, nothing changes until it changes inside you. Nothing changes in the world until you decide to see things differently, and take a risk as a result.

Sara Wachter-Boettcher doesn't just evangelize about ethical design in her books. On social media, she can be found amplifying underrepresented voices, critiquing the tech industry's blind spots, and building a supportive community that is inclusive of women, transgender, nonbinary, and queer people, as well as people of color. Sara is the principal at Rare Union, a product and content strategy consultancy, and the cohost, with Katel LeDû, of the podcast Strong Feelings, *which they describe as "a weekly dose of fun, feminist realtalk."*

CHAPTER 11

Someday Never Comes

Design for the Future Self

B y now, you've seen a wide range of techniques that designers can use to influence people's behavior. Yet, even with this extensive toolkit, it can still be hard to get results. Why?

If there is one major design challenge in changing behavior, it's asking people to make sacrifices today for the possibility of benefits sometime in the far future. Those benefits aren't guaranteed to accrue, and people may not even realize it if they do. How do you know that you've avoided getting cancer, or that your savings habit kept you from losing your house? And how are you supposed to keep big long-term goals top of mind when faced with hundreds of minor decisions every day? It's far too easy to focus on the here and now. Far-off goals competing with present-day demands are a recipe for behavior change failure.

The tactics you've learned along the way can help with the challenge of designing for the future self. Your job is to help people make small steps over time that eventually lead them to their future destination. You'll likely need to understand what makes new behaviors feel difficult for people and how to overcome them. Your users may look at where they stand today and see a far-off destination, but no reliable way to get between the two. Helping them define a meaningful purpose will illuminate a path. Along the way, you can create a choice architecture that guides goal-supportive behaviors and sprinkle in lots of social support from friends or even technology to help your users weather challenges. To keep users on the path, you must maintain their trust as they go. Designing for the future self, like most of the real-world behavior change design projects you'll encounter, requires recruiting multiple approaches to support users along their journey.

But first, you need to understand why the future self is such a tricky design target.

Why the Future Self Is So Far Away

Psychologically speaking, the future is far away. Most people logically understand that the person they'll be tomorrow is pretty much the person they are today, only a little older. But people are optimists. They often imagine a better future self and commit to things that their new and improved alter ego will tackle with aplomb. Then the alter ego fails to materialize.

That's why people make future commitments that don't appeal to them as activities for today. How many times have you seen a deadline approach and wondered why you agreed to take on that project? Or found yourself toeing the start line of a race and cursing the past version of you who thought you'd *want* to get up at 5 a.m. on a weekend?[1]

Often, the future activities that people commit to require them to have accomplished something by the time the activities arrive (think physical challenges, intellectual or artistic projects, or achieving a certain physical appearance before a social event). If people haven't thought through what that preparation entails, there's a strong possibility they won't be ready when the day comes. That can lead people to wrongly conclude they're not capable, when in fact they just didn't follow the right recipe to accomplish their goal. Unfortunately, these sorts of failures can dissuade people from future behavior change attempts.

Or, perhaps people focus on a goal, but postpone taking the steps to achieve it. If behavior change is going to be a drag, why not start next week? One of the cognitive biases that people have is the *present bias*, where they overvalue their experience right now relative to their experience in the future. Someone in the throes of this bias will delay behavior change as long as possible, maybe until it's too late.

People's tendency to misjudge their future wants and abilities leads to negative consequences, both large and small. It contributes to people canceling plans at the last minute when they realize that meeting up for coffee with an old coworker actually *doesn't* sound like fun now that it's today. It may sour people on the pursuit of goals when reaching them doesn't bring hoped-for levels of happiness. Most importantly for behavior change designers, people's inability to gauge their future needs presents major roadblocks to effective behavior change.

TIP LOOK TO THE PAST TO UNDERSTAND THE FUTURE

Steve Portigal noted in Chapter 5, "People are not good predictors of their future behavior. You can ask them what they're going to do, but you shouldn't believe it." That's why researchers often reach for behavior-based questions rather than questions about what a person *would* do in a scenario. How people acted in the past is a much better predictor of future behavior than their forecast.

---

1 These are my personal examples, but I'm sure you have your own version of the 5 a.m. weekend wakeup.

Behavior change design can help combat this persistent human error through a few steps. First, users may need help setting goals that have staying power and can be chunked out into concrete steps. Second, people who have a specific plan of action are far more likely to succeed, so behavior change designers can help by developing such a plan and gaining users' commitment to it. Finally, there are specific tactics that can help keep users engaged through long-term behavior change. Sprinkling them liberally into the action plan will help keep people on the path.

Set the Right Goals

The first step in designing for the future self is helping users select the right goals for themselves. These goals must have staying power if they're to anchor new behaviors over a long period of time. And as you now know, the most enduring goals are rooted in people's deepest values.

Remind Users What Matters

Giving people meaningful choices aligned with their personal values helps them stick with behavior change. Personal values don't tend to change very quickly over the course of a person's life. The things that are deeply meaningful to someone at 20 are very likely still meaningful at 60, even if the way they are expressed looks different.

When coaching people through behavior change, try to have them reflect on how their new routines will help them live into their values, like the Kumanu Purposeful app does. Purposeful asks users to describe their best self and then reflect on their day. How did their behaviors help make them more like who they want to be? The more that people are able to articulate these sorts of connections, the easier it becomes to bear some discomfort today in service of a future goal.

Dig Deeper

When someone doesn't want to try changing a behavior, it might be more about what they'd have to *stop* doing rather than what they'd have to *start* doing. Giving up a pattern of behaviors that's meeting some needs is a tough thing for people to do. So when someone seems reluctant to change a behavior, it's worth asking what benefits they're receiving from their current pattern—and seeing if there are other ways to provide those benefits.

NOTE JOBS TO BE DONE

There's a concept in service design called "jobs to be done" (JTBD). JTBD asks, "What job is your product hired to do?" It turns out, users often get a different benefit from products than their creators intended. If two people are fighting over an orange, it's important to know what jobs they want that orange to do. The solution to sharing the orange looks very different when the JTBD are "provide zest for a cake" and "use the juice for a mimosa" versus "eat the orange." If JBTD is already in your design toolkit, it can be repurposed nicely here.

Ask people what about their current behavior pattern works well for them. Sometimes the answer will help you design a different path to behavior change that honors what the person likes about their current routine. Someone who doesn't recycle because it's too much work might be willing to try a service that provides receptacles and picks them up for processing. Asking "why" gives you more material to work with in developing an action plan for behavior change.

Bring the Target Closer

The more distant a goal is, the more difficult it is for people to imagine it concretely. This means that accurate, realistic planning for how to reach a goal is much more difficult. Coupled with people's tendency to do what is easiest, a fuzzy future plan can mean no action is taken until it's too late to be effective. Incorporate near-term milestones into the goal-planning process so that users can set their sights on something that feels more real while working toward the far future. A way to get people thinking about the connection between now and later is to ask users to write down one thing they can do *today* to begin moving toward their goal.

Make a Plan to Reach Them

Once users have picked a goal that's appealing and enduring, it's time to sketch out the path for achieving it. Most meaningful outcomes take lots of work to achieve. That work may not be visible to someone who hasn't yet tried to change those behaviors. People see the new muscles, the triumphant purchase of the house, or the awarding of the diploma, without seeing the grunting in the gym, the vacations and new cars skipped, or the late nights in the library. It's your job as a behavior change designer to shine the light on that process and help users get comfortable with being uncomfortable.

Set Expectations

Recall from Chapter 4 that people who make informed decisions are more likely to stick with them when they get challenging. It's time to be informative! You can help people calibrate their expectations around behavior change to be more realistic. If the behavior change is one where progress is slow, like weight loss, be clear about what people can expect. And be honest. Lose It! warns new users that it can't provide an estimate of how long it will take them to reach their goal weight until they've logged at least a week's worth of food and exercise data (see Figure 11.1). While this may be mildly frustrating, it also sends a signal to users that they can trust the estimate they eventually receive.

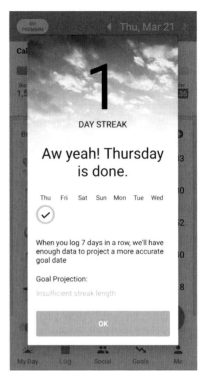

FIGURE 11.1

Lose It! will tell users approximately how long it should take them to reach their goal weight if they keep to the plan, but not until it has at least one week's worth of data to inform that estimate.

Create and Commit

Action planning is actually a behavior change technique in and of itself, but it's especially useful to help combat some of the inertia that people have with working toward a goal in the far future. The name does a good job describing how the technique works. After deciding on a goal, people detail the steps they'll have to take to achieve it. The action plan can be high level or very detailed, depending on what is most useful to the person. Good action plans also sometimes include decision rules to help people maintain their behaviors in challenging situations. For example, *if* I'm traveling for work and the hotel doesn't have a gym, *then* I'll do a bodyweight exercise in my room using an app.

Once an action plan exists, people need to make a *commitment* and pledge to complete the items in the plan. In its simplest form, a commitment can be a promise to oneself. New users of Pattern Health select behaviors from a list of recommendations and then sign their name to indicate their intentions to perform those behaviors (see Figure 11.2). Psychologically, the act of commitment helps to shift the behavior into a more autonomous form of motivation. That makes follow-through more likely. This is an example of the friction that Aline Holzwarth referred to in Chapter 4.

Another way to leverage the behavior change technique of commitment is to make it visible to other people. Whether it's the encouragement and support that other people provide when they know a friend is in pursuit of a goal, the guidance and advice that fellow change seekers offer, or just a reluctance to look foolish, making a public declaration of intentions may help people bridge the gap between behavior change today and goals for the future. A public commitment doesn't have to take the form of a social media post; telling selected friends and family will do. Encourage your users to share their goals with at least one other person who will help hold them accountable.

FIGURE 11.2
Pattern Health users select the behaviors they intend to perform from a personalized list of suggestions. After they've compiled their plan, they add a signature to signal their commitment.

Get Personal

Sometimes you might present people with an action plan that they don't care to commit to. There may be a real objection that's difficult or impossible to overcome. A low-sodium diet will be a bigger challenge for someone who cooks exclusively by microwave than for a gourmet chef. Testing blood sugar five times a day will terrify someone with a fear of needles. Increasing 401(k) contributions could be nearly impossible for someone barely meeting their basic necessities.

How can you help people in these situations change behavior if they are interested?

Don't get stuck on the *how* as much as the *what*. If there's any flexibility in the way that a particular goal can be accomplished, use it! What more palatable alternate routes exist to the outcome? Can you recommend low-sodium canned foods for your microwave cook? Coach the needle-phobic user on minimizing the pain of a finger prick? Help the person struggling financially to identify a support program that lightens the pressure a bit? While in general it's a best practice to streamline choices, when people say no to the first suggestion, it's a good idea to have some viable backup plans for them to consider.

Foot in the Door

Frequently, people are intimidated about committing to a big, hairy, audacious goal. Milestones can help wedge a foot in the behavior change door. Ask your user to commit to one of the early milestones in a bigger behavior change process rather than focusing on that long-term outcome. Milestones may be less scary for people to commit to. Plus, once users reach that first milestone, you can challenge them to carry on to the next. Eventually, people may gain the confidence to commit to the true end goal, or they may simply arrive there by having chased each consecutive milestone. Either way, it's a win.

Choose the Right Tactics

So what goes into a good action plan for a faraway goal? The *how* of the behavior change plan is as important as the goal it's designed to reach. Certain tactics specifically bridge the gap between immediate actions and faraway goals.

Highlight Small Achievements

People need to have faith that their efforts will build up to something meaningful. Help them key in to small signs of progress. No one magically changes overnight, but many behavior change efforts yield incremental changes over time. Some weight loss programs encourage people to take a weekly full-body photo of themselves in the same location. After a few weeks, the first and the most recent photo start to look different from one another, and can cement a person's belief in the process. When change is slow, small indicators of progress are important.

Ideally, you've helped your user select a goal that includes some milestones that can be reached more immediately. These milestones are a helpful way for users to see progress as they check them off the list. By their nature, people will be able to reach milestones much more quickly and easily than they can achieve the big goal. Acknowledge each milestone achievement. Use some of the tactics from Chapter 7 on using feedback to help people feel a sense of growth.

Shape the Environment

Deciding to pursue a big future goal may be a single decision, but in practice it shakes out to hundreds or thousands of smaller decisions. The person undertaking behavior change may have to decide multiple times a day, week, or month to take actions that move them toward the goal. As you learned in Chapter 4, this sort of repeated decision-making can wear on people. As their willpower wanes, the quality of their choices might, too.

You can reduce the need for this repetitive decision-making by structuring the environment to better support behavior change. This could mean removing prompts or cues for unwanted old habits, or facilitating new behaviors with reminders or tools. It could also mean equipping people with rules of thumb that change how they interact with the environment: I don't walk down the snacks aisle at the grocery store. Environmental improvements can reduce friction in performing tasks that might otherwise get procrastinated. Basically, they make it harder to prioritize the current self over the future self by moving the effort to the former rather than the latter.

Encourage your users to make changes to their physical environment to support their new behaviors. Eating healthy snacks is a lot easier if the pantry is potato chip–free. Your digital product can also provide some of the environmental shaping to support a behavior change goal. Whether it's pinging users with a reminder to record data or taking an action within the app (see Figure 11.3), enhancing their

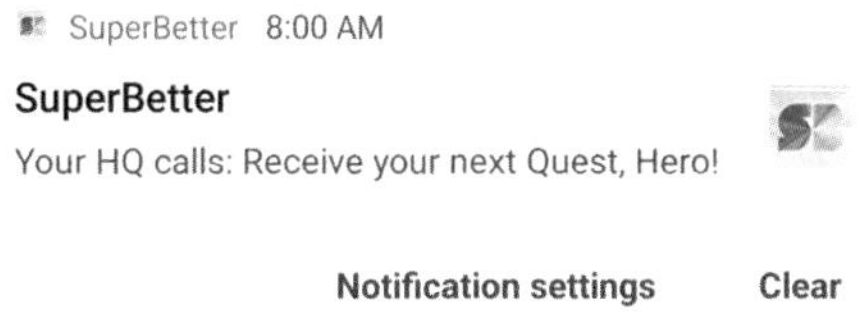

▮ SuperBetter   8:00 AM

SuperBetter
Your HQ calls: Receive your next Quest, Hero!

Notification settings   Clear

FIGURE 11.3

Push alerts like this one from *SuperBetter* can remind users to take app-based actions toward their behavior change goal.

ability to do a behavior through sound (see Figure 11.4), or having them automate a repeated behavior (see Figure 11.5), digital tools can also change the context in which the user acts.

Using environmental reshaping tactics reduces the amount of in-the-moment decision-making a person has to do. Your users make fewer but more effective commitments by spending their behavior change energy reconfiguring the environment to make the new behaviors seamless with their routines.

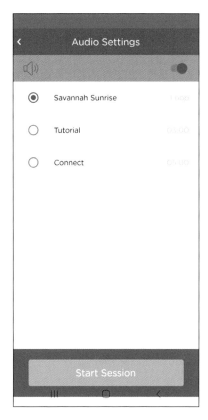

FIGURE 11.4
For behaviors like meditation, apps like HERE can support the new behavior with ambient noises like these soundtrack options.

FIGURE 11.5
Stash allows users to automate their investment contributions, so that they only have to decide and take action once in order to begin working toward a savings goal.

Get Mindful

Often, people opt out of behavior change because they value their comfort today more than the potential benefits of their efforts at a distant future time. Making the current day behavior change more rewarding may help some of these people change their minds about giving it a try.

Aside from offering incentives, you can make behavior change feel more rewarding by helping people notice and savor its short-term benefits. Although someone on a new exercise program might feel uncoordinated and sore, it's also very likely that they're experiencing a burst of energy after every workout. Can you help them notice that? Someone just beginning to save toward a goal may not have made much progress in an absolute sense, but they may have set themselves up for success by ditching a high interest credit card. Is there a way to show them how that small act will translate into dollars saved over time? These sorts of progress indicators feel good *now*.

One area of behavior change where designers can help orient people to near-term benefits is smoking cessation. Within days of the last cigarette, an ex-smoker's body begins to repair itself. In Figure 11.6, the EasyQuit app helps users recognize all of the physiological improvements they're experiencing six months after quitting.

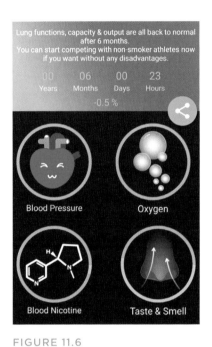

FIGURE 11.6

EasyQuit helps users pay attention to the many small ways in which their bodies work better six months after smoking their last cigarette.

For some types of behavior change, it is also important for people to learn to cope with discomfort for at least short periods of time. Easing them into the change, like Couch to 5k does with new runners, can serve double duty in keeping discomfort minimal while also helping people psychologically prepare to deal with it. Most behavior change journeys benefit from using discomfort productively, whether it's struggling with a new type of math problem or making a more prudent financial choice with an eye toward a bigger goal. How can you help your users view difficulties as a sign of progress and not obstacles? Connecting discomfort to meaningful goals can make it more bearable.

Bring the Future to Life

One tactic to make the future self more emotionally real is to picture it, literally. Some technologies are able to create realistic visualizations of what people might look like decades in the future, using images of their current appearance. The technology can manipulate the image to reflect different possible future selves depending on today's lifestyle behaviors.

Seeing how smoking or a sedentary lifestyle might impact their appearance can help people make changes now. At least one study using a high-quality 3D imaging technology from Medical Avatar (see Figure 11.7) found that seeing "future self" images helped people in a weight loss program make more progress. Another set of studies found that people made better financial decisions after looking at digitally aged photos of themselves; seeing their older faces helped them make the future self a higher priority now. The realistic imagery closes the emotional gap between people's present and future selves, which makes it harder to prioritize one to the other's detriment.

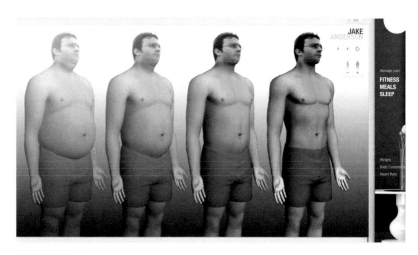

FIGURE 11.7

Medical Avatar's technology produces realistic versions of how a person might look in the future given a particular set of behaviors.

Not all future self images are created equal. Figure 11.8 shows what the Smoking Time Machine app thinks I will look like in 20 years if I continue to smoke. They didn't ask how much or how often I smoke, and offered only one option (smoking yes/no) to influence my appearance. The effective future self visualizations seem to work because they're credible; I'd be surprised if this one, which makes me look like a Halloween decoration, were effective.

Most likely, you don't have the technology to show people manipulated images of what their future selves might look like. That's okay. Similar effects have been found by asking people to vividly imagine their future selves, or having them complete activities that build empathy with their future selves like writing a letter to themselves. They may not be as memorable and fun as an aged selfie, but these techniques can still help bridge the gap between today and tomorrow, and they're much easier to implement.

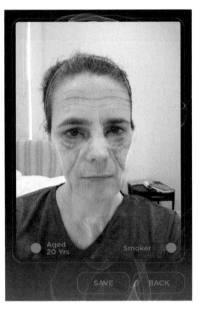

FIGURE 11.8

The Smoking Time Machine ages your selfie to show what you'll look like in 20 years as a smoker. The more realistic a future self simulation is, the more effective at helping to change behavior.

Know When to Fold 'Em

Sometimes, people decide that the changes needed to achieve a result aren't ones they want to make. It's not worth it to them to give up something that plays a positive role in their life, even if doing so provides a benefit. You may know a smoker, for example, who is fully aware of the health effects of cigarettes but gets so much stress relief and enjoyment from them that they're willing to take the risk of illness or early death. A lot of people would rather continue to enjoy the pleasures of "bad" behaviors than sacrifice them for an ambiguous future good. These sorts of cost-benefit analyses are totally logical, and they're significant barriers to behavior change.

You can always plant the seeds of change. But if your target user is truly not interested, the best strategy may be to walk away. If someone has decided that this particular behavior change journey is not for them, all you can do is let it go. A graceful goodbye[2] leaves the door open if the person changes their mind later; trying to force the issue, on the other hand, may drive people away forever.

The Upshot: Make the Future Feel Real

It's challenging to get people to commit to big future goals that require making small sacrifices now. Behavior change designers can help people choose and successfully pursue far-off goals. Working with users to pick big goals that align with their values, and breaking those goals into meaningful milestones can make behavior change more appealing. Translating nebulous goals into concrete action plans also helps; users may be overwhelmed with figuring out a path to success, but willing to commit to a clear plan.

Getting people to work today for the benefit of their future self requires a combination of tactics to be included in that action plan. You can help people chart out a course of action that they believe will succeed, which may mean providing evidence that the steps will work. Make it easier to connect emotionally with the future self, who, after all, is just a later model of today's self. And show people how to find the positives in behavior change so that the efforts today don't feel as much of a slog.

---

2 Much like the graceful goodbye you say to users who quit your product.

*Kate Wolin, ScD, is a serial digital health entrepreneur whose weight manage-
ment program, ScaleDown, was acquired by Anthem in 2017. Here are some of
her tips for keeping people on track with behavior change over time.*

Why is it important to show users progress quickly?

When we were creating ScaleDown, the science consistently showed
that if people didn't lose *some* weight in the first 10 to 14 days of a
weight loss program, they were highly unlikely to achieve the ultimate
goal of 6% weight loss in six months, which is what the gold standard
programs achieved.

How can you keep users engaged with near-term feedback?

The very first physical activity interventions in the commercial space
were fitness wearables, and only focused on one or two metrics of
physical activity like step count. You weren't noticing the ancillary
benefits and feelings that a coach might have focused on. We have to
focus our messaging on what people will gain from the experience, and
not just what they're going to give up. That's why a lot of health behavior
interventions that researchers or scientists design tend to have people
monitor more than one thing. It might be that being active today will
help you sleep better and wake up tomorrow feeling more rested.

What if someone is not ready to change now?

Someone might not be ready today to engage today, but they might be
tomorrow. This always stood out when we were trying to engage cancer
patients. What we would hear was, "I am totally overwhelmed right now.
I can't think of putting anything else on my plate other than my treatment." No one would argue with that. But just because someone says
no today, doesn't mean the answer's going to be no tomorrow because
circumstances change. One of the really unique opportunities that
technology has put in front of us is this ability to be able to identify the
right time to offer a program to someone.

When people are engaging in weight loss, they hit a goal, and typically
it's, "Okay, good job, you're on your own now. Bye, bye," and people
regain the weight. So at ScaleDown, we would say, "People who lose
weight and successfully keep it off long term continue to engage in
high-frequency, daily self-weighing. But we recognize you may not want
or need daily feedback from us anymore, so we're going to ratchet down
the intensity, and put you into maintenance mode." The intent of that
was that relapses happen for all behavior change. We know this, and
it's so normal that we actually build it into behavior change curricula.
When you have a maintenance mode that allows you to stay connected

to someone in some way, when life events happen that lead users to cope with stress with their friends Ben and Jerry instead of their buddies down at the gym, then it allows the platform to trigger a message that says, "Hey, do you want to flip back into weight loss mode?"

A lot of times traditional behavior change programs have this "all or nothing" approach. We haven't really used technology to generate this in-between "Hey, let's stay connected so that if the time comes that you need a little help, we're still here. It's not a failure. It's actually so normal and so expected that we'd love to hang around and be here for you." If we reach out to people, they feel really cared for as a person. When we connected with people as humans in the past, we might have had that conversation, and it's been a little bit lost in the design of technology products.

 *Kate Wolin, ScD, is a behavioral scientist and epidemiologist bridging the academic and commercial worlds. Since the successful exit of her startup, ScaleDown, Kate has spent time as the Chief Science Officer of Interactive Health, an Adjunct Associate Professor at Northwestern University, and consulting on the development of digital health solutions as the cofounder of Coeus Health. Kate's research focuses on cancer prevention and survival, and the role of physical activity and obesity in oncology care. She is an expert in translating evidence-based science to digital interventions.*

CHAPTER 12

Nothing's Gonna Stop Us Now

Go Forth and Engage

This is the end of the book. You've now covered behavior change design for digital products 101. I've covered a lot of ground and provided a broad overview of the many ways that behavior change can be included in design. Like a 101 course, this book is intended to be the start and not the end of your education on the topic. I hope what you've learned here has been interesting enough that you'll pursue more knowledge.

In this book, I've focused primarily on how psychology can influence the design of digital products. I've also tried to point out specific places where nondigital experiences can support what's happening on the screen and how behavior change design needs to account for offline behaviors. As you go forth and use behavior change design, you'll certainly need to adapt some of the tools in this book to fit your specific situation. The modality of your product should and will influence how behavior change principles get expressed. The tools at your disposal will present opportunities and limitations for what you're able to do with behavior change. Think of it like a puzzle, not a problem. This is a fun part of the challenge!

Doing Behavior Change Design

The quickest and most effective way to learn behavior change design is to *do* behavior change design—with an asterisk. I've emphasized a focus on outcomes throughout this book. That same focus is necessary for your practice with behavior change design to translate to learning. If you're trying new approaches to product design but not measuring their effects, you don't have the information you need to determine whether your tests worked or not. When you're using behavior change design, formulate a clear goal at the outset of your project.

For a beginner behavior change designer, some learning-oriented goals might be:

- If I include questions about capability, opportunity, and motivation during my user research, does it help me with developing product requirements?

- When I build my product around user goals and provide psychology-based feedback for users, how often do they actually achieve those goals?

- Are users engaging more often or more deeply with my product when I use behavior change design principles to design it?

- Am I making different decisions about how to design my product when I think about user motivation? What do those differences look like?

Answering these questions may be easier if you compare them to other projects you've worked on where you didn't use behavior change design.

Getting others on your team to buy into behavior change design as a way of working might be a challenge. The best way to convert nonbelievers is to show them a success story. One option is to point to a case study from another company who has used behavior change design to positive effect (and even better if it's a competitor—no one wants to be behind the times in their industry). Your outcomes logic map, from Chapter 2, can also come in handy. The leading indicators are often ones that companies monitor anyway, so if you can track them in the service of more meaningful outcomes, it's a great way to show success.

Be Glinda, Not Elphaba

Behavior change design can be used for good as well as evil. You've seen examples of both throughout this book, and many of the experts I interviewed brought up abuses of behavior change during our conversations. Two things are often true when behavior change is used to coerce users into doing something they might not have chosen freely:

- The product designers are focused on business goals over user goals.

- The product designers either have not thought about the ethics of their choices, believe that what they're doing is right, or are so enchanted with their product that they don't think that someone else could be plotting how to exploit it.

People inside the companies misusing principles of engagement aren't evil;[1] they're just so myopically focused on their own needs that they don't realize they're not universal. It's like the witches in *The Wizard of Oz*. In the classic movie, Glinda the Good Witch is the foil to the Wicked Witch of the West. In the 1995 update *Wicked*, the

---

1 Usually. True villains aren't as common as the news would have you believe.

reader learns that the Wicked Witch, Elphaba, is more misunderstood than evil. Regardless of her intentions, the effects are the same. Elphaba's misdeeds set off a chain of events that result in her death.

Don't be like Elphaba.

I believe that being educated in behavior change design means having an ethical obligation to be alert to potential misuse of the science and to speak up if you detect it. You may not be the ultimate decision-maker about what's included in a product or how it's deployed. Putting truly unethical cases aside, in that case, the best you may be able to do is to voice your concerns, surface any evidence you have, and advocate for the choices that best support user needs. As Sheryl Cababa said in her interview in Chapter 6, it's not always realistic to walk away from a job with high ick potential, but raising concerns might motivate others to support a change of course. It may also plant the seeds for a company to reconsider its direction sooner rather than later if—when—their misdeeds are caught out.

I'm not trying to be a wet blanket by mentioning the ethics of behavior change design throughout this book. The reality is, there are always bad actors somewhere in the system, and if design teams aren't vigilant about guarding against them, users will suffer. As technologies like artificial intelligence, machine learning, and facial recognition mature, it will be important that designers think carefully about when and how to leverage them. If these technologies are used, design teams should have a plan to protect their users against the potential for data misuse.

Keep your eyes open. Ask the hard questions, imagine the worst-case scenario, and act in good faith.

Be a Behavior Change Evangelist

It's my hope that you bring some of the lessons of this book into your own work practice in a way that inspires your colleagues to want to do the same. Aside from buying copies of this book for everyone on your team, one way you can evangelize behavior change design is by talking openly about how you're using it in your work. Let others know that you're trying a new approach and invite them to observe and collaborate. If you try a technique from the behavior change toolkit and really like it, share that story so others can try it too.

As you tackle meatier challenges within your products, you may find that you need to bring in the big behavior change guns. Sometimes a subject matter expert is needed to really get an intervention right. In my experience, this is especially true when dealing with complicated health conditions or specialized populations; no amount of primary research from a nonexpert will get the product in shape as quickly as consultation with an expert.[2]

You may also find yourself faced with engagement challenges that aren't related to the topic of your intervention. For example, you might want to have a randomized control trial of your product to prove its efficacy versus a competitor's, but you don't have the knowledge or resources to set one up. External collaborations to the rescue! You can hire an individual expert or a consultancy that does behavior change work, or reach out to a local university. Academic researchers often have the know-how and resources to run a study, but not the access to a commercial quality digital product to test. A collaboration there could be a win-win.

Fueling the Competence Engine

Like all people, you have a basic psychological need for competence. For behavior change design to become a regular part of your practice, you'll need to see your skills improve over time. Of course, one of the best ways to get better at anything is to do it, so find as many opportunities as you can to bring elements of behavior change design to your work. Not everything you try will be successful, but every attempt is an opportunity to learn. Be thoughtful about when and why you're using a behavior change technique, and think critically afterward about why it did or didn't work. Being deliberate will make practice more powerful. You can also seek out education for yourself to continue expanding your toolkit.

Never Stop Reading

Because it's so new, there aren't that many books that blend behavior change and design. But there are tons of fabulous books about behavior change and design separately that will educate you and spark new

---

2 It's outside of the scope of this book, but participatory design is an excellent tool to employ in these situations. With participatory design, the people who will use your solution are invited to cocreate potential solutions during the design process so they can lend their firsthand experience and insights.

ideas. Plus, many individual behavior change designers and consultancies working in the space publish blog posts, white papers, and case studies. Organizations like the American Psychological Association often publish magazine-style summaries of the latest science on issues like mental health, parenting, and the environment. These can be a great way to quickly get up to speed on a topic for a project.

If you're comfortable reading primary journal articles, subscribe to receive the tables of contents from your favorite journals when they're published. This is an easy way to scan titles and keep abreast of new studies related to your work. Or set up a Google alert for keywords related to your area of interest so that you get an email whenever something on the topic hits the news.

You can also become a better behavior change designer by reading entirely outside the field (something Sara Wachter-Boettcher also suggested in her interview). Fiction builds empathy for people different than yourself. History teaches lessons about how past technologies were introduced and used. Science and science fiction offer different lenses of looking at the world and an invitation to imagine alternatives. Immerse yourself in other perspectives and information, and it can only help your practice.

Listen to Podcasts

Podcasting is entering something of a golden age. There are dozens of design- or behavior change–focused podcasts that feature expert guests. If you have a commute where you can listen to a podcast (or if you're one of those strange people who likes to exercise to podcasts instead of Robyn and Rihanna), some excellent ones to subscribe to are the following:

- *99% Invisible* by Radiotopia
- *Action Design Radio* by the Action Design Network
- *Behavioral Grooves* by Kurt Nelson, Ph.D., and Tim Houlihan
- *The Bettercast* by Habitry
- *Design Better* by InVision
- Freakonomics Radio by WNYC
- *Hidden Brain* by NPR
- *Human Tech* by The Team W

- *Real World Behavioural Science* by the Behavioural Science and Public Health Network

- *Why'd You Push That Button?* by The Verge and the Vox Media Podcast Network

For the most part, these podcasts explore specific issues in behavioral science and design through interviews with experts. Not every episode of every podcast may feel relevant, but often there are nuggets of wisdom to be found and treasured. Similar to reading widely, listening to podcasts on topics that are ancillary to your main interests can improve your knowledge more than you might expect.

Talk to People

When you can, get information directly from the horse's mouth! Attend interesting talks at conferences and events, especially if the speakers are people you haven't met yet or the topic is aslant of your main work focus. Use networking sessions to thank the speaker for their perspective and ask a specific question or two; having an agenda for the conversation makes approaching a stranger feel far less awkward, and most people appreciate a compliment on their presentation.

Beyond that, reach out online to people who are doing interesting work. Join regularly scheduled Twitter chats using industry hashtags—you can find these with a little searching—and pose follow-up questions to people who intrigue you. There are also groups on other social platforms such as Meetup, Slack, and Facebook where you can chat with others who are working in behavior change design. Join and use them!

Even reaching out via email is fine, although I strongly recommend having a defined request that does *not* include the phrase "pick your brain." Some of the things you can ask for that are far less annoying include copies of a person's paper that you'd like to read, insight on a specific question relevant to the person's expertise, or for them to speak to your team about their work. You may not get a response to cold outreach, but you increase the odds if you have a precise request.

And finally, talk to your peers and teammates about behavior change and how it might be applied in your work. Share insights and questions with each other. Workshop ideas. Start a club and take turns picking a behavior change book to read and discuss. Talking with other new behavior change designers can sharpen your thinking and help you get to the next level.

Above all, never stop learning. Behavior change, like any science, is in constant motion. New research is published daily, and sometimes old research is discarded. Meanwhile, the potential applications of behavior change also evolve as technology matures. With a field that perpetually grows, behavior change designers are always *becoming*. The learning is never done.

The Promise of Behavior Change

The real appeal of behavior change design is its potential to put people—users—permanently on a more positive life path. That may seem like hyperbole, but when people are ready for a change, the right intervention can transform their lives. People can become emotionally invested in behavior change without being attached to any one tool to achieve it. As both Vic Strecher and Josh Clark pointed out, it's a victory when behavior change designers inspire their users to move on from their tools to carry out their new behaviors in the world. To achieve this, think about how to design a sticky *process* more so than a sticky *product*.

All of the tactics in this book are designed to help shift people's reasons for behavior change into the identified ("I do this because it helps me achieve important goals") or integrated ("I do this because it's part of who I am") motivational types. These sorts of motivation are long-lasting and resistant to obstacles. They can be the engine for forever change.

Realistically, your users won't use your product forever. Technology will render your app obsolete. Users who access it through their employer will change jobs. Or perhaps your users will "graduate" as they master the level of behavior change in your product. But if the products that people use when they are behavior change beginners help them invest in the process, then they'll be equipped to find the right new tools or supports as they move throughout life.

Ideally, your behavior change intervention is like the training wheels on a bike. They allow people to take their first ride without fear of falling. Over time, riders gain the confidence to pedal faster and the skill to balance without help. The training wheels become an encumbrance instead of a benefit. If all goes well, people will remove the training wheels. But they won't stop riding. Their journey has just begun.

Kate Lawrence builds great teams. Kate's ability to identify talented people and her approach of coaching and empowerment provides a model to follow in bringing a new discipline to your team. And Kate's track record of democratizing the UX research function in her organizations offers lessons for integrating behavior change design into any product development process. When it came time to get an expert perspective on how to build a new competency of behavior change design within teams and organizations that haven't previously tried it, I could think of no better person to ask than Kate.

How can you get other teams to participate in a new capability?

It's not just about creating training and tooling and platforms and opportunities for people to learn how to conduct their own research. It is about the shoulder-to-shoulder curation and this concierge process of going through it together. There will be pieces you feel empowered to do by yourself, but which are very supported by the experts on your team. The journey of incorporating user feedback without support just doesn't work.

We learned that people want to be guided shoulder-to-shoulder as they try to bring forth the voice of the user. It's our job as researchers to recommend specific tools, platforms, and methods for the question at hand. For example, we can empower people by partnering to formulate a question guide, uploading it to a platform for conducting the research, setting up an online panel, and then enabling teams to self-serve research. They have all of the recordings from the research and can begin analyzing the findings with our support. It's empowerment plus the added insurance of our team's oversight, so we can step in and help elevate the findings to strategic insights. Bringing in the appropriate tooling and support scales the reach of the research team.

Creating opportunities for people to witness and hear users' feedback is powerful; it is an experience that resonates. You come away owning a piece of it. Developers have commented, "Now that I've seen the [user] challenges and pain points firsthand, I can't un-see it, I can't un-hear it." It makes the work of solving those user problems feel more personal and more urgent.

How do you get others to include a new capability in their process?

In my experience, the way to grow a research capability is simply, to say yes—to every request, at least initially. Each positive research experience generates great insights, goodwill, and positive word-of-mouth, and that will help the group gain traction. Sometimes during those early days, it can feel like you are drowning! But saying yes to projects means saying yes to users, and that is critical during the nascent phase of growing a research discipline.

Every single conversation you're having, try to turn it into a research opportunity. I coach the researchers on my team to always start with "yes," even though the initial request may need some finessing. When someone comes to us and says, "I need a card sort," or "I need a usability test"—it's possible that the better method in that case might be a survey or key user interviews. But starting with yes, and then an exploration of the most suitable method, is better than saying "no" and trying to change someone's mind at the outset. Being collaborative and agreeable up front allows the conversation to continue, and when you gain someone's trust, that's when you have an opportunity to guide them to a different method or a different timeline.

Sometimes people aren't open to suggestions, and that's still workable. In those cases, I recommend that we do a version of what they are requesting, and then supplement with a sample of what we recommend to show the potential. That's why I love hybrid methodologies and the mixed method multimethod approach, allowing for capturing user feedback through more than one lens.

How does a focus on insights keep research at the center?

Findings are generated from individual studies, and are smaller and more localized than insights, which tend to be larger, more evergreen, and more actionable. As researchers, we are in the business of uncovering findings and then synthesizing these into strategic insights that we deliver to the business. The value of research is helping to guide decisions about granular elements, such as detail of a page navigation, but also keeping an eye on the larger experience of the product or app. Meeting users' expectations happens within a product or app and also in sync with the experience that users have in their larger digital ecosystem.

Kate Lawrence is UX Research Director at Akamai Technologies, where she's building a UX research function and democratizing access to research insights across the organization. Previously, Kate helmed the User Research and Design team at EBSCO Information Services. Kate's prior companies also include SilverRail Technologies, World Travel Holdings, and TripAdvisor.

INDEX

Better Business Bureau, 240

between subjects analysis, 27

biases, cognitive, 72–73, 82–84, 107, 193, 255

Bickmore, Timothy, 244–245

biofeedback programs, 9

Biogen, 197

blockers. *See* ability blockers

BlockSite, 195

boredom, 149–151

broaden and build theory of positive emotions, 242–243

Build Better Products (Klein), 25

Busuu, 175–176

C

Cababa, Sheryl, 141–142, 272

Call of Duty video game, 152–154

Cambridge Analytica, 237

cancellations, 249–250, 265–266

cancer survivors, teens, 214, 218–220

capability (C)

  in COM-B model, 98–99

  intervention functions for, 121

  physical and psychological, as ability blockers, 99–102

  in research, 277–278

caregiver arrangements, 241

CarePath, 147

cartoonish avatars, 244–245

case studies, 33

Castro Sweet, Cynthia, 39–40

celebrity endorsements, 234

census data, 251

Centers for Medicare & Medicaid Services, 167

certificates, 240

Change Talk, 56–57

chatbots, 217–221, 242. *See also* Vivibot; Woebot

checklists, 79–80, 109–110

choice architectures, for easier decisions, 74–89

  appealing choices, 86–87

  cognitive biases, working with, 82–84

  constraining the choice, 75–76

  friction and fuel, 91–92

  level of detail, 77–78

  red face test, 88

  self-care, 82, 83

  social desirability, 74, 88–89

  structure with checklists, 79–80

  symmetry, 76–77

  trade-off awareness, 80–82

  using fun, 84–86

choices, meaningful, 41–64

  alignment of choices, 55–62. *See also* alignment of choices

  autonomy and, 13, 42

  core values, what really matters, 45

  designing by the rules, 51–55. *See also* options for users

  designing for user autonomy, 47–48

  hooking users, 48–51

  life purpose, 63–64

  ownership, importance of, 42–44

  understanding the decision, 48

  want to do rather than have to do, 46–47

exposure, 24–26, 29

extended social networks, 186

external motivation, 12

Eyal, Nir (regret test), 88

F

Fabulous app, 76, 86–87, 209

Facebook

  Messenger chatbot, 218

  newsfeed algorithm, 246

  personalized preferences, 207

  personalized recommendations, 208

  social interaction in design of, 221

  user data sharing, 236–237

Farmville game, 46

FDIC, 240

fear of missing out (FOMO), 194

feedback, 151–166

  badging, 167–168

  biofeedback programs, 9

  for children, 163

  creating motivational, 151–152

  cumulative, 156–157

  frequency of, 162–163

  gamification, 165–166

  good, negative, and positive, 152, 157

  immediate, 156–159

  IRB review for puzzle study, 36

  measurements, 154–156

  mistakes in, 166–168

  multiple levels of, 156–159

  normative, 156–157, 159–162, 242

  purposes of, 154, 168

  for repeat achievers, 163–165

  types of, 154

  user control and expert success, 247

financial services. *See also* Stash

  algorithms and user control, 246–247

  finance management, 176, 213

  interventions, 4

  long-term outcomes, 23

  retirement savings programs, 135, 136, 230

Fitbit, 89, 155, 164–165, 166, 190, 206

fixed mindsets, 144–145

flow, 148–149

fly in the urinal, 155

fMRI image of brain, 212

focus, as ability blocker, 95

formal accountability buddies, 179–180

formal insights research, 96–98

Foursquare, 167–168

free text options, 220

freemium model, 235

future commitments, 255

future self, design for, 253–268

  action plans to reach goals, 257–260. *See also* action plan for future self

  goal setting, 256–257

  reasons for so far away, 254–256

  sustaining the journey, 267–268

  tactics to reach goals, 260–265

  user exits, 265–266

opportunity. *See also* physical opportunity; social opportunity

ability blockers, 103–105

in COM-B model, 98–99

intervention functions for, 121

overcoming blockers, 125–127

opportunity costs, in choices, 71

opt outs, 52–53, 135–136

opting-in *vs.* opting-out, 134–136

options for users, 51–55

maybe later, 53–54

multiple choices, 54, 220

no choice, 54–55, 74

opt outs, 52–53, 134–136

organ donations, 135

outcome bias, 73

outcomes logic map, 21–22

outcomes story, 20–29

analysis plan, 27–29

baseline, 27

identification of behavior changes, 23–24

long-term success metrics, 22–23

measuring exposure, 24–26

specific data, 26–27

P

Pacifica, 42–43

Pagoto, Sherry, 193

pain management, 167, 233

paradox of choice, 67–68

participatory design, 273

Pattern Health, 91, 259

Pavlok, 129

peer-reviewed journals, 35–36, 39

peer support, 178–179, 186

Peloton, 197–198

performance management interventions, 5

permissions, 40, 237–238

personal best, 190

personal information

privacy policies and legalese, 238–240

security of, 221, 228–229, 236–237, 249

personalizations, 206–212

for ability, 136–138

coaching, 211–212

data-driven, 210–211

for future self, 259–260

preferences, 206–207

recommendations, 208–210, 246, 247

personas, 9

perspectives

building toward value, 17–18

of choice architecture, 91–92

intervention functions and systems lens, 141–142

life purpose, 63–64

pragmatic scientific rigor, 39–40

research capabilities, 277–278

robot therapists, 225–226

social support and sharing, 200–201

sustaining for future self, 267–268

thinking differently about trust, 251–252

user research on barriers to behavior change, 113–114

workplace performance management, 169–171

persuasion, as intervention function, 120, 121, 128, 130

Pet Rock, 204

pets, 104–105, 204

photos, digitally aged, 264–265

physical capability, 99–100, 121–122

physical opportunity, 103–105, 111, 126–127

piano stairs in subway station, 84–85

Picky Eater Project, 47

Pinterest, 240

playlists, 209

podcasts, 274–275

Pokémon Go!, 5

Pollyanna principle, 249

pop-ups, and value propositions, 61–62

Portigal, Steve, 113–114, 255

positive feedback, 152, 157

Practical Empathy (Young), 106

practice, 130

pragmatic scientific rigor, 39–40

preferences, personalized, 206–207

prescription, in behavior change design, 15

present bias, 73, 255

priorities, as reflective motivation, 111

priority scores in data grid, 117–119, 134

privacy policies, 238–240

proactive communication, 222–223

product research, 35–37

product success story. *See* outcomes story

productivity hack, 131

progress, and growth, 144

progress bars and indicators, 242, 263

psychological capability

as ability blocker, 100–102, 111, 118

solving as ability blocker, 122–125, 132, 137

psychological needs, 13–14

punishments, in motivation, 107, 128–129

Purposeful app, 63, 256

Q

question laddering, 45

R

Radiotopia, 274

randomization, 31

randomized control trials (RCTs), 30–32, 273

reactive communication, 222

readiness levels, 53–54

real-world comparisons, 32–33

recency effect bias, 73, 249

recommendations, personalized, 208–210, 246, 247

red face test, 88

reflective motivation, 99, 105, 111

solutions to blockers, 128–129

regret test (Nir Eyal), 88

relatedness, as psychological need, 14

reminders, 131

research

academic research, 35–36

building new capabilities, 277–278

finding ability blockers, 96–97, 109–112

product research, 35–37

unexpected discoveries, 113–114

ACKNOWLEDGMENTS

When I first learned I'd be able to write this book, I made a few giddy posts on my social media accounts thinking a few members of my family and some close friends might care. I wasn't prepared for how supportive my extended network was, and how many people would reach out with congratulations and offers to help. This process has been a touching reminder of what a wonderful group of people I've met along my career and how so many of them have contributed in ways large and small to the content of this book.

I can point to a few inflection points in my life that shaped my career path and this book. The late Nalini Ambady and Chris Peterson both nudged me to take big life steps; I hope they would be proud of where I ended up. Working at HealthMedia was a game changer; I don't think I've ever learned so much so fast or laughed so hard. I was fortunate while at HealthMedia/J&J to have Sean Badger and Rich Bedrosian as managers and mentors for a killer product-psychology combo. Cass Sunstein planted the crazy idea of writing a real book in my head, and gave me the confidence to pursue it. And in my current role at Mad*Pow, I'm supported by an incredibly talented team. Amy Heymans, Mike Hawley, and Dustin DiTommaso deserve particular thanks for carving out space for behavior change design to grow as a discipline and me to grow as a practitioner.

I owe an enormous thank you to the people who agreed to be inter-viewed for each chapter. I admire every one of them and feel honored to include their perspectives in this book. Heather Cole-Lewis, Cynthia Castro Sweet, Vic Strecher, Aline Holzwarth, Steve Portigal, Sheryl Cababa, Diana Deibel, Josh Clark, Alison Darcy, Sara Wachter-Boettcher, Kate Wolin, and Kate Lawrence; how did I get so lucky?

Jeff Kreisler, thank you for bringing your unique blend of smart and funny to the foreword. I owe you one (or three).

I was also asked to identify technical reviewers to comb through the manuscript for factual accuracy. Against my own self-preservation instincts, I tapped five of the smartest people I know, and am grateful for their thorough and helpful feedback. Emily Azari, Omar Ganai, Rob Gifford, Mike Ryan, and Steve Schwartz, thank you.

Similarly, I am thankful to the people who took the time to write testimonials. I didn't realize before writing a book that the author has to *ask people for compliments*. How horrifying for a lifelong sufferer of impostor syndrome! Thankfully a stellar crew came through: Alyssa Boehm, Jen Cardello, Nir Eyal, Kim Goodwin, Amy Heymans, Lis Pardi, Scott Sonenshein, and Matt Wallaert. Y'all made me blush.

The team at Rosenfeld Media is world-class. Marta Justak, thank you for your editorial prowess. Sometimes you were pretty blunt with me, but you were (usually) right. You made my ideas so much better. Lou Rosenfeld, thank you for taking so many months of back-and-forth to help shape my initial proposal into something worth writing, and for your support along the way.

Finally, thanks to my friends and family who helped make sure I was fed, watered, and occasionally played with over the long months of writing this book. More importantly, thank you for believing in me and my ability to bring this book into the world.

 **Rosenfeld®**

Dear Reader,

Thanks very much for purchasing this book. There's a story behind it and every product we create at Rosenfeld Media.

Since the early 1990s, I've been a User Experience consultant, conference presenter, workshop instructor, and author. (I'm probably best-known for having cowritten *Information Architecture for the Web and Beyond*.) In each of these roles, I've been frustrated by the missed opportunities to apply UX principles and practices.

I started Rosenfeld Media in 2005 with the goal of publishing books whose design and development showed that a publisher could practice what it preached. Since then, we've expanded into producing industry-leading conferences and workshops. In all cases, UX has helped us create better, more successful products—just as you would expect. From employing user research to drive the design of our books and conference programs, to working closely with our conference speakers on their talks, to caring deeply about customer service, we practice what we preach every day.

Please visit rosenfeldmedia.com to learn more about our **conferences**, **workshops**, **free communities**, and **other great resources** that we've made for you. And send your ideas, suggestions, and concerns my way: louis@rosenfeldmedia.com

I'd love to hear from you, and I hope you enjoy the book!

Lou Rosenfeld,
Publisher

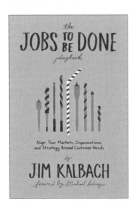

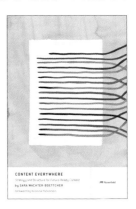

ABOUT THE AUTHOR

 **Amy Bucher, Ph.D.,** works as a Vice President of Behavior Change Design at Mad*Pow, a purpose-driven strategic design agency in Boston. Amy crafts engaging and motivating experiences that help people change behaviors that contribute to physical, mental, and financial health and well-being. This involves planning and conducting research and translating insights into strategy and requirements for products and end-to-end experiences spanning digital and real-world components.

Previously, Amy worked on behavior change products in-house at CVS Health, Johnson & Johnson, and HealthMedia, and has prior healthcare industry agency experience working for Big Communications on an innovation team. Amy received her A.B. from Harvard University, and her M.A. and Ph.D. in organizational psychology from the University of Michigan, Ann Arbor.

Amy tweets at @amybphd and when she's not spending her writing energy on a book, blogs at amybphd.com.